Residential QCI Handbook

BEYOND THE NREL JTA

Paul H. Raymer

Salty Air Publishing

Falmouth, MA

Paul H. Raymer/Salty Air Publishing
157 Palmer Ave.
Falmouth, MA 02540
www.saltyairpublishing.com
www.HeyokaSolutions.com

Ordering Information:
Quantity sales. Special discounts are available on quantity purchases by corporations, associations, and others. For details, contact the "Special Sales Department" at the address above.

Residential QCI Handbook/Paul H. Raymer. —1st ed..
ISBN 978-0-9906781-3-7

Dedicated to all the incredible people in the building science community who dedicate their lives to making other people's lives safer, more comfortable, and more energy efficient.

If all you have is a hammer, everything looks like a nail.

—Proverb & Law of the Instrument

Quality Control (Assurance) Inspector

Even with the best intentions, mistakes get made. A Quality Control Inspector moves the result of a project one step closer to perfection. To do that, the inspector has to be a generalist and understand all the aspects of the project and have the experience of making his or her own mistakes and have learned from them. The weatherization of a home has many parts from an initial analysis to determine what needs to be done, to understanding the systemic nature of construction, to appreciating the needs and resources of the homeowner, to the capabilities of the crew performing the work, to the verification that the results match the initial expectations.

A Quality Control Inspector is a residential energy efficiency professional who ensures the completion, appropriateness, and quality of energy upgrade work by conducting a methodical audit/inspection of the building, performing safety and diagnostic tests, and observing the work.

Imagine that there is a small house owned by a nice old lady, we'll call Edna, who is struggling to meet her bills and tolerating exceptionally cold conditions in the winter and excess heat in the summer, conditions that make her life miserable. The house fits into the local weatherization program and a BPI (Building Performance Institute) certified energy auditor has visited the house and created a work order to make improvements. He did a blower door test to measure the leakage, measured the insulation depth in the attic, determined the existing insulation in the walls, and tested the atmospherically vented combustion, gas fired water heater, furnace and oven for safety. And the crew comes in and begins to work.

The weatherization crew consists of a certified Crew Leader and a good crew who performs consistently good work. You are the quality control inspector on the job and during an in-progress inspection you find that the auditor mis-identified the building envelope and the installers are not insulating a wall between the conditioned and unconditioned space, a living room and enclosed porch. They are insulating the walls of the enclosed but unconditioned porch. They will complete wall insulation this afternoon, according to their schedule. If you stop the job, you will miss the completion time and extend the job. If you don't stop the job, will you be doing your job as the quality control inspector?

During the site visit, you interview Edna to see if she is satisfied with the work. During this interview, she expresses concern that the crew has tracked dirt through her house. The interview itself is a 'soft skill', getting to know the client and understanding her point of view. How do you satisfy the client but not give the crew a bad reputation? How do you verify what the client is telling you?

During the final inspection your blower door reading shows you the house needs additional air sealing. To help identify where air-sealing efforts should focus, you conduct zonal pressure diagnostics, starting with the house to attic pressure. With the house at 50 Pascals WRT (With Reference To) outside, you measure 40 Pascals in the house WRT the attic. What do you do with that information? What do you tell the crew leader about missed air sealing opportunities?

You are beginning to wonder about both the auditor and the crew leader and then you determine the attic hatch was not installed according to program requirements and looks very sloppy. The client complains. This is not the kind of work you expect from this crew. The homeowner now has serious doubts about what she is getting. Her husband had built the house, and she didn't want it altered in the first place, but her son talked her into it and now she really doubts that she has done the right thing.

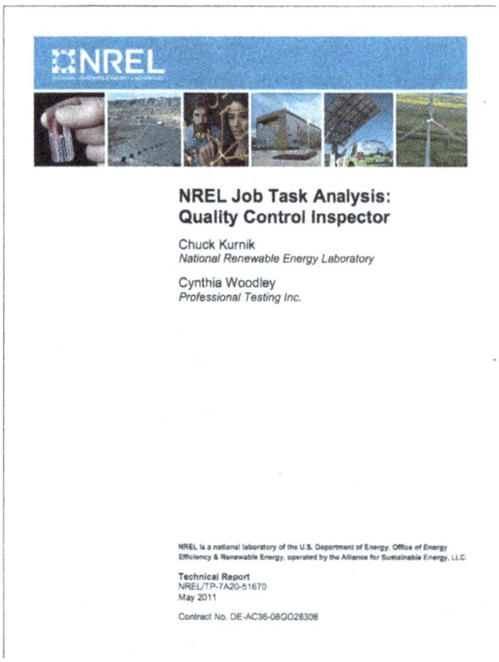
NREL

NREL Job Task Analysis: Quality Control Inspector

Chuck Kurnik
National Renewable Energy Laboratory

Cynthia Woodley
Professional Testing Inc.

NREL is a national laboratory of the U.S. Department of Energy, Office of Energy Efficiency & Renewable Energy, operated by the Alliance for Sustainable Energy, LLC.

Technical Report
NREL/TP-7A20-51670
May 2011

Contract No. DE-AC36-08GO28308

The audit for the home indicates the un-floored attic was previously uninsulated, and the work order increasing the insulation to R-38 insulation throughout the entire attic, which adheres to your program requirements. During final inspection you discover that about half of the attic had existing R-13 batt insulation, which the installers left in place and blew over with another R-38 of cellulose. What would be the most appropriate way to address this? She does have more insulation, but how could the auditor miss this? Did he actually bother to stick his head up in the attic and look?

These are typical challenges that a Quality Control Inspector faces every day. It requires knowing building science well enough to guide an experienced crew leader or crew around mistakes that could not only cost the client and the program money, but could also jeopardize the life of the building and the occupants. The Quality Control Inspector is the final authority, the place where the buck stops. The QCI has to know the rules of the program he/she is working in as well as the applicable codes. The QCI has to understand and have an eye for job site safety. The QCI has to understand the technology of the tools and what they can and can't say. And most difficult of all, the QCI has to be able to listen. He/she has to be able to know when to deliver a tough message with a commanding voice and how to deliver a consoling message to cool down a hot head.

Bare in mind that Quality *Control* is based on Quality *Assurance*: assuring that the quality has been included in the work. It's not an easy job, but when all is said and done, it is a very satisfying job that helps a lot of people individually, collectively, and for the world and our grandchildren.

How to use this book

If you have 4 pencils and 7 apples, how many pancakes will fit on the roof?

1. Purple, because aliens don't wear hats.
2. All of the above
3. None of the above
4. Above the above
5. It depends

You may feel that some of the questions you face on multiple choice tests are like that. But don't get hung up on the quality of the question. Stay true to what you know.

The structure of this book is based on the DOE NREL (Department of Energy National Renewable Energy Laboratory) Job Task Analysis (JTA) for Quality Control Inspector (QCI). This is the same structure that the IREC (Interstate Renewable Energy Council) accredited training programs use to build their courses and the same structure that the WAP website (http://1.usa.gov/1abEQbz) courses are based on. (The complete contents of the NREL JTA can be found on the web at http://1.usa.gov/1tRT7gM).

The QCI JTA has five Domains broken down into individual tasks. Those tasks are broken down further into individual bullet points. The field exam is based on all of those bullet points. It is important to note that this is a *national* certification; not a local one.

You should be able to practice your skills and knowledge anywhere in the country. Maybe you don't experience a whole lot of Category I, atmospherically vented water heaters in your part of the world, you still have to know how to test them to pass this exam.

This book amplifies those domains, tasks, and bullet points going through them in a logical order. Not everything happens on cue when you are engaged in your field exam. For example, there might not be an opportunity to identify a potential fall accident issue. However, you need to let the proctor know that you know when and where fall protection gear would be appropriate. You should take care to mention every item on the Field Guide, which is every item that is part of the Field Test. (QCI Field Guide is located at http://bit.ly/1I44PgG).

The multiple choice certification exam is a closed book, two and one half hour exam with 100 questions. The field or practical exam is three and one half hours long. (Exceptions can be requested for additional time.)

Residential building science is not limited to low-income weatherization projects. Although this book is based on the structure that the DOE has created for that program, the role of a quality control inspector spans the entire spectrum of construction. Sloppy workmanship is a long-term nightmare. The most expensive issues for contractors are callbacks. Callbacks account for 1 to 2 percent of sales, as much as $500 per home! Doing it right the first time is much less expensive. It is vitally important that the crew doing the work appreciates what they are doing, that they are invested in the success of doing it right. Weatherization can be a particularly dirty, tedious, frustrating job and the pay is commonly low. It has to be something more than money that serves as motivation. That attitude comes down from the leaders of the project - particularly the Quality Control Inspector. Perhaps they should be called "Quality Control Guides".

At the end of each section I have listed the activities that are a part of the BPI/QCI field test, things that you should know. Note that I have listed them where I believe they are appropriate to the Domain and Task and so some activities may be listed more than once.

As Columbo the TV detective used to say, "Just one more thing." I want to emphasize that I am not a supporter of paperwork for the sake of paperwork. There is no need to keep a daily inventory of the paperclips in your desk. Good paperwork is designed to make life simpler; not more complicated and documentation is vital. It should save time and it should save trouble. But it should also be considered just as much a part of a job as doing the job itself. Taking a tool out of a case, using it, and returning it to the case are all part of the job and all are equally important. I am a great fan of checklists because they keep me from forgetting things, they provide an overview of what is left to be done, and they provide documentation that each step was done. However, there is a balance between a useful checklist and a checklist of checklists!

A brief note about Standards: The BPI Building Analyst Standard was created a long time ago and is the Standard that the QCI certification is based on at the time of this writing. The BPI/ANSI 1200-S-2015 is the replacement Standard. However, indications from BPI are that after January 1, 2016 the original BA Standard will continue to be the reference standard for the QCI exams. The 1100 and 1200 Standards will continue to apply to "Skilled Building Performance Certifications" (i.e. BA, Envelope, Heating). Both Standards are mentioned in this book.

Who am I to tell you all this and why should you listen to me?

I have a degree in creative writing. I never went to school to study building science (of course, when I went to school there wasn't such a subject). I am not an engineer. What I have learned about houses I have learned by living in them and working on them and working with people who have worked on them. Houses are a magical conglomeration of materials that somehow all work together in one way or another to shelter and comfort and house the people who live in them. From caves to mansions their occupants make them unique. If the walls could talk, every one of them would have stories to tell - some boring, some violent, but many loving and caring. And they all slowly drift back toward the earth from which they started in different forms and need to be adjusted and maintained to perform the way we want them to.

Since I started making a living working with houses in 1977, the technology has changed a great deal. We have much more sophisticated materials and tools than we had before. It is virtually impossible to build a house that is so loose that it doesn't need mechanical ventilation. When I started, I was told to measure infiltration by the number of quarters that would fit between the window sashes! Codes and standards have changed to demand better energy performance. There are over a hundred million single family homes in the US and a lot of them are poorly built and more continue to be built, which means that I can guarantee that there will be work improving and fixing and repairing houses forever.

Although I have spent a lot of time working with them, I have also never had the honor of working full time with a weatherization agency improving older homes. I have the greatest respect for people who have the knowledge, skill and attitude to crawl around under the bellies of mobile homes or into the deep, dark, sweltering recesses of attic crawl spaces. I just want to mention that I am overwhelmed by the quality, skill, dedication, and attitude of the multitude of trainers and administrators who dedicate their lives to others. This is not an easy or glamorous profession. Getting the job right, day after day, house after house, is a challenge to the mind and the body. It is easy enough to stand in front of an audience in a clean, tempered, well-lit classroom and show pictures of these places that my building science mentor, John Tooley, calls Hell. It is quite another thing to go there day after day.

So this book is filled with information that has worked for me, and although I have spent considerable time in the deep, dark recesses of homes, this book is not based on the daily routines of a construction company or weatherization agency. Undoubtedly you are going to be subject to your organization's systems and checklists and work orders. You're going to have to filter what I have written by what you have to work with every day. But I hope these words help you to clarify some things. I hope you will always keep learning because there is always more to know, and we can always do an even better job.

Paul H. Raymer
IREC Certified QCI Master Trainer
Falmouth, MA

May, 2015

Just a note about training videos. The electronic version of this book was created with additional interview and discussion videos from fourteen extraordinary national trainers whose wisdom amplifies its contents. The trainers include:

Chris Baker – Training and Technical Coordinator for FSL Southwest Building Science Training Center;
James Corriera – Technical consultant for the State of Connecticut Weatherization Program;
Chris Clay – Senior Trainer for the Building Performance Center;
John Dolen – MSW, psychologist;
Amanda Hatherly – Program Manager for the New Mexico EnergySmart Academy;
Phil Hull – Director of Training for Community Housing Partners;
Joshua Jacobs – Principal of Certified Building Analysis;
Rick Karg – President of RJ Karg Associates and Residential Energy Dynamics (RED);
David MacLellan – Program Manager at CLEAResult;
Matt Redmond – Director of Training for New York State Weatherization Directors' Association (NYSWDA);
Adam Romano – Director of Training Operations at the Association for Energy Affordability;
Bill Spohn – President of TruTech Tools;
A. Tamasin Sterner – President and Chief Coach at Pure Energy Coach, LLC;
John Vaughn – Program Manager at Cape Save, Inc.

The videos are available from www.HeyokaSolutions.com

NREL Quality Control Inspector JTA Outline[1]

Quality Control Inspector Specifications and Content Outline

Job Description: A Quality Control Inspector is an evaluator who verifies the work performed against the work plan, specifications and standards, performs building diagnostics, records/reports findings and concerns, and specifies corrective actions; by conducting a methodological audit/inspection of the building, performing safety and diagnostic tests, and by observing the retrofit work; in order to ensure the completion, appropriateness and quality of the work providing for the safety, comfort, and energy savings of the building occupants.

Domains/Tasks	
Domain 1:	**Conducting Quality Checks - In-Process Visual/Sensory Inspections**
Task 1:	Verify worker compliance with safety rules
Task 2:	Assure worker professionalism
Task 3:	Address work problems
Task 4:	Evaluate client satisfaction regarding the in-process work
Domain II:	**Conducting Quality Checks - Post-Work Visual/Sensory Inspection**
Task 1:	Review client file and the work scope
Task 2:	Perform an exterior and interior visual/sensory inspection
Task 3:	Evaluate client satisfaction
Task 4:	Determine pass/fail of the work
Domain III:	**Conducting Quality Checks - Post-Work Diagnostic Inspections**
Task 1:	Conduct health and safety tests
Task 2:	Conduct diagnostic tests
Task 3:	Identify work problems
Domain IV:	**Ensuring Worker Professionalism**
Task 1:	Perform spot checks
Task 2:	Provide feedback regarding professionalism
Domain V:	**Ensuring Program or Project Compliance**
Task 1:	Maintain professional credentials
Task 2:	Confirm the allocation of public/private funds
Task 3:	Evaluate installed measures against the field guide, SWS and state/local codes
Task 4:	Close out the project
Task 5:	Maintain files and records

[1] This is the Specification and Content Outline from the NREL JTA. Note that Domain 1 uses the number 1 and not the Roman Numeral as the other Domains do. I have maintained that convention.

Domain 1: Conducting Quality Checks

KSA - Knowledge, Skills, and . . . Attitude

Conducting Quality Checks and In-Process Visual/Sensory Inspections requires the ability to observe what is going on on the job site - observe the workers, and the site conditions, interview the crew chief and the client, and work in confined spaces. It requires knowledge of basic construction techniques, codes and standards adopted by local jurisdictions, Federal regulations, first aid, the use and application of diagnostic tools, and how to interview people and find out what they are thinking.

It requires skill in communication, observation, and remaining tactful. It also requires practiced skills in the application of the diagnostic tools - not just how to use them but skill in applying the information that they provide. There is an assumption that if a digital display on a piece of expensive equipment tells you something, it must be right. And it probably is . . . as long as the operator set it up correctly. Diagnostic tools need to be told the parameters that they are working from: for example, how big is the hole in the blower door? If the manometer doesn't know what ring is installed, it can't calculate the airflow. Which model blower door are you using? What kind of fuel is the boiler burning?

Beyond the operational basics, most of these diagnostic tools can provide more information. A QCI needs to know more than just how to turn it on or off and get a number. A QCI needs to know what the number means and know when it is wrong!

Then there is Attitude. Sometimes the A in KSA is interpreted as Ability. Someone might be Able to climb a mountain but that doesn't mean they have the required Attitude to make it to the top. Attitude is a combination of self-confidence, will power, desire almost to the point of lust, and personal feedback when the task is done well without the need to brag. A QCI reflects that attitude back at the crew leader and the crew as well as the homeowner. Attitude can't be taught - but it can be learned.

Task 1.1 Verify Compliance with Safety Rules

The first task in the first domain is "Verify worker compliance with safety rules". To accomplish that you will need to know what the safety rules are. Although having received OSHA 10 or OSHA 30 training is not one of the prerequisites to take the QCI exam, it would be a good idea to have one of those credentials under your belt since OSHA (the Occupational Safety and Health Administration) makes the rules for safety and does the job site inspections. If you don't know what the rules are, how are you going to comply? The JTA does require knowledge of Federal Regulations like OSHA and EPA. You can take an OSHA 10 (10 hours) or OSHA 30 (30 hours) course on-line or in a classroom. (https://www.osha.gov/dte/)

OSHA is focused on safety and there are a lot of rules. The JTA doesn't go into safety anywhere close to what OSHA is looking for, but the fact is that if you are doing construction work you need to comply with OSHA's rules. An injured worker will not be good for the well being of your project to any degree. You need to know what you can and what you can't do. I once asked for a Band Aid on a government project, and got no response to several requests. I finally wrapped the cut up in toilet paper and duct tape. I was told later that if they had provided a Band Aid they would have had to fill out a stack of forms describing what happened.

But you're going to want to look for obvious safety issues. Simply having a safety checklist can help you to prove to your proctor that you know what to look for. You might want to use a state checklist and modify it to suit your particular situation. There are dozens of free residential construction safety checklists on the Internet. (The one included with this book is from Michigan.) Safety is a serious issue, not just for the health of workers but also for the health of a business or program. Lawsuits are time consuming and expensive. You don't want to go there.

Where is the first aid kit? Is it up to date and how do you know? Where are the SDS (MSDS) sheets for the materials that are being used? Are they up to date and how do you know? Are there confined spaces on the site and is there a confined space entry procedure? Are unprotected sides or edges more than 6' in height guarded? Are personal fall arrest system components inspected prior to each use? Is there a log?

What about Personal Protective Equipment (PPE)? Are workers wearing their eye protection, hardhats, hearing protection, safety-warning vests? Is there a person on the site competently trained and assigned the task of being in charge of safety? Is that person skilled in first aid?

The best trick I ever learned was from a property manager, who suggested that we apply a thin layer of petroleum jelly to our upper lip before entering certain units to mask the odor.

Yes, the work of the QCI is no more glamorous than the Auditor, and in fact could be seen as far more demanding, time consuming and intrusive to the resident who has already been tortured by installers.

Having a bunch of guys traipsing around your house in PPE is not comforting and always provokes questions. I know from personal experience that my patience wears very thin, when I am working in adverse conditions. I also know how good it feels to get outside and clean up. I routinely utilize wet wipes or wash and Dry towelettes before I get back into my personal vehicle.

James Correira: Technical Consultant for the State of Connecticut Weatherization Program

During the job site walk around, you need to point out safety issues and how they are being addressed. All of the issues that are on your safety checklist might not be applicable to the project you are using for your field test. No one can be an expert in everything, but know where to get the information you need when you need it.

Lead Safety

One of the documents that should be in the client file is the Pre-Renovation Form. It is a confirmation sheet signed by the occupants that they have received a copy of the lead hazard information pamphlet about the potential risks of lead hazard exposure from renovation activity. They should sign this lead safety form before work begins. http://www2.epa.gov/sites/production/files/documents/pre-renovationform.pdf

BPI QCI Field Test

- Identify measures that would have required fall protection;
- Identify appropriate respirator requirements in the scope of work;
- Identify appropriate PPE in relation to work in confined spaces;
- Identify appropriate PPE in relation to the general work requirements of the work scope;
- Identify potential jobsite hazards.

1.1.1 Walk Around the Job Site

A job site may have just one crew of workers who know each other well and have worked together for a long time. Or it may be various workers or crews with varying qualifications and tasks - plumbers, electricians, carpenters, and insulation installers - working compatibly together. Add to that the homeowner whose house is being invaded and daily routines impacted, who may be fearful for the safety of his or her possessions. And then the inspectors walk in, including the Quality Control Inspector who has the power to stop or change the work, criticize or praise the workers, soothe or exasperate the homeowner. Do you walk onto the site with a necktie, shiny shoes, and a clipboard? Or tons of makeup, a short skirt, high heels, and a cell phone to your ear? The terms in your title are scary: Quality . . . Control . . . Inspector. "Who is this person and how are they going to ruin my day?"

That's both a good thing and a bad thing (not ruining a day but being somewhat scary). Somehow you need to get across that all of the people working on this house are working together to accomplish a common goal: to end up with the most energy efficient home for the client. You respect the work that is being done and you understand that there are a lot of different and usually complex tasks and a lot of different people with varied levels of skills and knowledge working together and mistakes get made, issues get overlooked, and houses are dynamic objects - things change! Your job as the Quality Control Inspector is like the Conductor of the Orchestra getting all of the professional musicians (including the concert master or crew leader) to play together and make beautiful music or, in this case, a comfortable and energy efficient home.

One of the keys to making this happen is your skill in communication. How you present yourself on the job site is part of that communication – arrive appropriately dressed and equipped, organized, efficient, and focused on this particular project. You can't have a cell phone strapped to your ear or be tapping out text messages. You need to know how to deliver your information, but almost more importantly, you need to be able to listen to what the homeowner, crew chief, and workers are telling you. You can't effectively generate quality without being able to listen.

Is the job site safe? You need to have a working knowledge of what "SAFE" means and how it is defined by OSHA and other federal, state, and local jurisdictions. Does the job site look clean and neat? Are there extension cords stretched across the floor where people can trip on them? Are the furnishings of the house protected from debris, dropped tools, and other accidents? Are the workers wearing the necessary PPE (Personal Protective Equipment)? Are you? How's your knowledge of First Aid? When were you certified? Where is the First Aid Kit? Has anyone checked it recently to make sure that it is fully stocked and up to date? Where is the nearest Emergency Room? Where would you get access to up-to-date SDS (MSDS) information for all the materials that are being used on the project?

The Crew Chief should be able to walk you through most of those issues. But what if he doesn't? You need to know how to listen to him/her. It's possible that no one has ever asked those questions before.

While you are walking around a job site, you can start putting together a punch-list of things that need to be completed. If your organization doesn't have a stock punch-list, you could make your own, download one from the internet, or use this Punch List (http://bit.ly/1LV4ldX).

There is a difference between a punch-list and a checklist. Which one you use depends on your role in the project. You might use a punch-list to list the outstanding items that need to be completed in the project if you are directly associated with the process of the project.

A project checklist might be more than that. As a QCI you might be engaged to score a project for another organization, to determine whether or not the contractor has successfully completed the required goals. Consistency is the key there. A contractor needs to know that the houses that he/she works on are going to be scored consistently and not randomly at the whim of the QCI.

The following screen shot is page one of a two page checklist that was developed by A. Tamasin Sterner of Pure Energy Coach to provide consistent scoring of projects that they verify. Each item has a click box that adds to the total score. If forms are missing, that's 1 point. If mechanical ventilation is not vented to the outside, that's 10 points. A total score of 10 points may mean that the house fails the inspection depending on the program. Contact Pure Energy Coach http://www.pureenergycoach.com/for more information.

QC INSPECTION CHECKLIST

PURE ENERGY QUALITY ASSURANCE

DATE	JOB #	PROGRAM	ELECTRIC COMPANY	GAS COMPANY
QC INSPECTOR Dyba		W	WPP	COL GAS
INSTALLATION CONTRACTOR DCW		W+	MET-ED	
CUSTOMER NAME		WW		

CONSUMER EDUCATION ACHIEVED BY CONTRACTOR ☐ YES ☐ NO ☐ UNDETERMINED

0 TOTAL SCORE

RESULT

- 10 or MORE ☐ FAIL
- 5 - 9 ☐ PASSED WITH ACTION REQUIRED
- 1 - 4 ☐ PASSED WITH COMMENTS
- 0 ☐ PASSED WITH NO RESERVATIONS

PROCEDURE COMPLIANCE

NOTES:

10	Excessive moisture and/or mold inside the pressure barrier and air sealing and/or insulation was done	5	IAQ too poor for any air sealing to be done
10	Priority list not followed	5	Property not returned to pre-existing (or better) condition
10	Invoiced item not found	1	Completed file not uploaded
5	Test data missing	1	Forms missing
5	Decision-making documentation incomplete	1	Forms illegible

HEALTH AND SAFETY

NOTES:

10	Undiluted CO in the water heater flue gasses exceeds 100 ppm and the inspection was within 30 days of the invoice submittal date		
10	Undiluted CO in the boiler flue gasses exceeds 100 ppm and the inspection was within 30 days of the invoice submittal date	10	Mechanical ventilation not vented to the outdoors
10	Undiluted CO in the furnace flue gasses exceeds 100 ppm and the inspection was within 30 days of the invoice submittal date	10	Flue and/or chimney problems
10	CO in range flue gasses exceeds 100 ppm and the inspection was within 30 days of the invoice submittal date	5	Water heater spilled flue gasses in excess of 5 minutes and the inspection was after 30 days of the invoice submittal date — worst case and natur
10	Gas leak(s) found and the inspection was within 30 days of the invoice submittal date	5	Boiler spilled flue gasses in excess of 5 minutes and the inspection was after 30 days of the invoice submittal date — worst case and natur
10	Gas leak(s) found and the inspection was within 30 days of the invoice submittal date and gas leak(s) now fixed	5	Furnace spilled flue gasses in excess of 5 minutes and the inspection was after 30 days of the invoice submittal date — worst case and natur
10	Flame rollout observed and the inspection was within 30 days of the invoice submittal date	5	Water heater draft pressure below the minimum and the inspection was after 30 days of the invoice submittal date — worst case and natur
10	Ambient CO exceeds 35 ppm and the inspection was within 30 days of the invoice submittal date	5	Boiler draft pressure below the minimum and the inspection was after 30 days of the invoice submittal date — worst case and natur
10	Insulation is in contact with non-IC rated heat producing fixture	5	Furnace draft pressure below the minimum and the inspection was after 30 days of the invoice submittal date — worst case and natur
10	Insulation is in contact with flue or chimney	5	Undiluted CO in the water heater flue gasses exceeds 100 ppm and the inspection was after 30 days of the invoice submittal date
10	Water heater spilled flue gasses in excess of 5 minutes and the inspection was within 30 days of the invoice submittal date — worst case and natur	5	Undiluted CO in the boiler flue gasses exceeds 100 ppm and the inspection was after 30 days of the invoice submittal date
10	Boiler spilled flue gasses in excess of 5 minutes and the inspection was within 30 days of the invoice submittal date — worst case and natur	5	Undiluted CO in the furnace flue gasses exceeds 100 ppm and the inspection was after 30 days of the invoice submittal date
10	Furnace spilled flue gasses in excess of 5 minutes and the inspection was within 30 days of the invoice submittal date — worst case and natur	5	CO in range flue gasses exceeds 100 ppm and the inspection was after 30 days of the invoice submittal date
10	Water heater draft pressure below the minimum and the inspection was within 30 days of the invoice submittal date — worst case and natur	5	Gas leak(s) found and the inspection was after 30 days of the invoice submittal date
10	Boiler draft pressure below the minimum and the inspection was within 30 days of the invoice submittal date — worst case and natur	5	Gas leak found and the inspection was after 30 days of the invoice submittal date and gas leak(s) now fixed
10	Furnace draft pressure below the minimum and the inspection was within 30 days of the invoice submittal date — worst case and natur	5	Flame rollout observed and the inspection was after 30 days of the invoice submittal date
10	Ground cover not installed properly	5	Dryer venting material improper
10	CAZ Depressurization exceeds limits	1	CO alarm doesn't meet material or installation requirements

8/14

1.1.2 Knowledge of Safety Issues

Basic Construction Knowledge: You wouldn't have gotten this far without knowing a great deal about how houses go together. You know the names of all the basic component parts so you wouldn't tell someone to, "Go over there and grab that thingy by the whatchamacallit and do that . . . you know . . . to whatever!" You speak the language, talk the talk, and have experience with the construction trades from the basement to the attic, from plumbing to insulation, from the bottom chord of a truss to the sill plate.

Weatherization: A thorough familiarity with the WAP guidelines is important: Title 10 Part 440. http://bit.ly/1PzJwFH

Particularly note Appendix A Part 440 - Standards for Weatherization Materials. Some familiarity with the financing issues would also be helpful: Title 10 Chapter II, Subchapter H, Part 600 (http://bit.ly/1crKZ3S). Chris Baker of SWBTC (Southwest Building Science Training Center) recommends at least a read through of Part 600 just in case some issues come up on the written test.

Codes & Standards Adopted by local jurisdictions: And you probably have some familiarity with the local codes and standards. It is important that you stay up-to-date on these as much as possible. It is a moving target. Your local jurisdiction may or may not have a Building Inspector. There are a few resources that you can refer to:

International Code Council: "The International Code (http://www.iccsafe.org/) is a member-focused association. It is dedicated to developing model codes and standards used in the design, build, and compliance process to construct safe, sustainable, affordable and resilient structures. Most U.S. communities and many global markets choose the International Codes." (http://bit.ly/1FVvjBX)

This is the organization that puts all the basic building codes together. States commonly base their codes on the 'Icodes', most of which change every three years. It is a member-based organization so if you have an idea of how the codes can be made better, you can send in that change and go to the meetings, and impact the national codes.

Public Resource.org: (https://public.resource.org/

Copies of the codes and standards for free! The codes of each state are listed and you can download them. This is an extraordinary resource.

State Energy Efficiency Codes: http://www.energycodes.gov/adoption/states

Department of Energy Status of State Energy Code Adoption - the official website for energy code adoption by state.

National Fire Prevention Association: http://www.nfpa.org/codes-and-standards

"The mission of the international nonprofit NFPA, established in 1896, is to reduce the worldwide burden of fire and other hazards on the quality of life by providing and advocating consensus codes and standards, research, training, and education.

The world's leading advocate of fire prevention and an authoritative source on public safety, NFPA develops, publishes, and disseminates more than 300 consensus codes and standards intended to minimize the possibility and effects of fire and other risks."

Building Codes Assistance Project: http://bcap-energy.org/

"BCAP strives to be the premier resource for energy code support, coordination, technical assistance, news and information. Our mission is to reduce the energy consumed in the construction and operations of buildings by working with national, state, and local governments and other stakeholders to promote the adoption and implementation of building codes and standards."

BCAP provides a service with the acronym OCEAN (http://bcap-energy.org/ocean/) Online Code Environment and Advocacy Network. You can click on the map of the country and the state you are working in and get the up-to-date information on the latest codes that are in force - a snapshot of everything you need to know about the codes.

Database for State Incentives for Renewable Energy: http://www.dsireusa.org/

This is a resource for incentives and funding for renewable energy projects. It is not focused on codes, but it is a great resource for getting the work done.

OSHA: https://www.osha.gov/# The Occupational Safety and Health Administration, a part of the Department of Labor, should be familiar to you. It is the Federal Agency that regulates the health and safety of workers. There are OSHA 10 and OSHA 30 hour training sessions. Familiarity with that training will allow you, as a Quality Control Inspector, to note issues on the job site that need to be addressed. Just imagine that you are an OSHA inspector. What would you do?

First Aid: It is an unfortunate fact of life that accidents do happen on job sites. It is a sort of macho-thing that issues like first aid get looked upon as a sort of namby-pamby activity. Just suck it up and get on with the job. But when something happens, it happens quickly. Decisions have to be made quickly and decisions require knowledge and preparedness. Knowing where the first aid kit is and that it is fully stocked as well as knowing that there is someone on the job site who is trained to use what's in the first aid kit will go a long way in improving the timeliness of a response to an accident from a small abrasion to a heart attack.

What's in your First Aid Kit? When was the last time you checked the dates on the contents? A job site first aid kit is different from the kit that you have at home. As the QCI you need to check with the Crew Leader about the contents of the kit. Your agency or company should have a policy for updating, replacing, and renewing the contents. There should be a log as to when this last happened. The crew leader and at least one other person on the job site should know the location of the kit. And at least one person on the job site must be trained in first aid including CPR.

Based on Federal OSHA Regulation Standard 1910.151.b: "Adequate first aid supplies shall be readily available." This rule applies to treatment of minor injuries that occur in the workplace.

Every business needs to meet the standards set forth by OSHA, as well as local and state labor laws... A well designed OSHA Compliant First Aid kit will meet and exceed the OSHA Standards, as well as serving the needs for the target group for which that first aid kit was designed... based on years of study, accident and injury reports, and first aid logs, kits need to meet or exceed OSHA standards, and be designed with respect for American Medical Association (AMA) guidelines.

http://www.first-aid-product.com/ofinterest/osha_ansi.htm

A typical 105-piece, 25-person bulk kit that meets all OSHA and ANSI standards, and is designed for use in small offices, vehicles and work sites would consist of at least the following components. The case should be partitioned and water-resistant, be wall-mountable, and be designed for quick transport.

Kit should include:

- (20) Fabric bandages, 1" x 3"
- (16) Adhesive plastic bandages, 1" x 3"
- (3) Knuckle fabric bandages
- (2) Fingertip fabric bandages

- (1) Triangular sling/bandage, w/2 safety pins, 40"
- (1) Conforming gauze roll bandage, 2"
- (1) Conforming gauze roll bandage, 3"
- (6) Gauze dressing pads, 3" x 3"
- (4) Gauze dressing pads, 4" x 4"
- (1) Trauma pad, 5" x 9"
- (2) Sterile eye pads
- (6) Alcohol cleansing pads
- (12) BZK antiseptic towelettes
- (6) Antibiotic ointment packs
- (6) Insect sting relief pads
- (6) First aid/burn cream, 0.9 gm. packs
- (1) Eye wash, 4 oz.
- (1) Instant cold compress, 4" x 5"
- (6) Aspirin tablets
- (1) First aid tape roll, 1" x 5 yd.
- (1) Kit scissors, angled blades, 4"
- (1) Tweezers, plastic, 4"
- (4) Exam quality vinyl gloves, 2 pairs
- (1) First aid guide

Red Cross: http://rdcrss.org/1E0LH2t

The QCI field test doesn't require you to ask about the location or contents of the first aid kit, but it sure wouldn't hurt to mention it.

Standard Work Specifications: (https://sws.nrel.gov/) The story goes that when the US got into World War II there was an immediate need for war materials - guns, bullets, tanks, ships, airplanes, etc. and they all had to work and most of the experienced people had been sent off to fight. A whole new, efficient work force had to be trained in a hurry. Standardized work procedures had to be developed that focused on performing a functional and reliable series of steps.

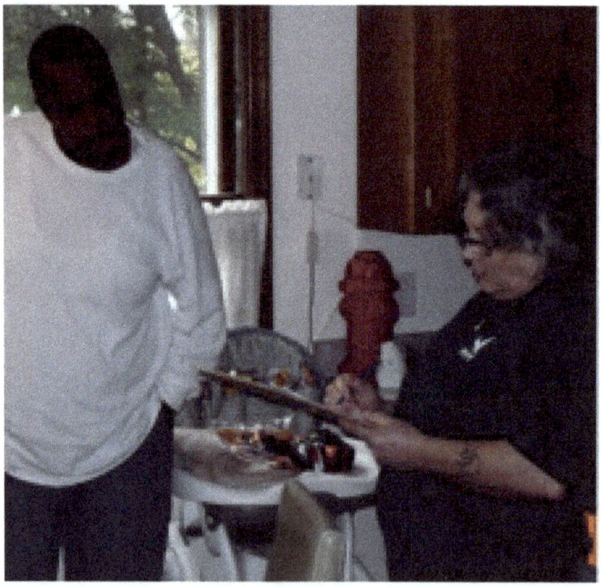

Much more recently, under the leadership of Vice President Biden, the Department of Energy realized that with over one hundred million single family homes in the US, there was a need for skilled people to weatherize them. The houses were there. The materials were there. The tools were there, but there weren't enough people with enough knowledge to get the job done. That brought about the creation of the Standard Work Specifications or SWS. The procedures in the SWS are fundamental they include a description of how to perform a task and a description of why you are doing it. They are not state of the art, but if everyone follows at least these basic steps, the task will be successful.

The SWS are a fundamental part of the QCI exam. You should sign up on the SWS site and get familiar with how to find stuff. It's a great reference beyond the QCI exam. DOE (http://1.usa.gov/1FmUWrC)

You need to ask the crew leader enough questions to verify that he/she has thought through the details of the project. Asking about the location of the first aid kit and the SDS (MSDS) information is a simple check of important details.

1.1.3 Interview Techniques

As the QCI you don't need to perform the in depth client interview (although an abbreviated one may be on the field test). The energy auditor should have done that, and you should have the records as part of the client file. The auditor should have established rapport with the client, settled down at the kitchen table or some such common place, and found out what the client knows about the house, what needs to be improved or fixed, if there are any areas that are out of bounds for the crews, changes to the house, health concerns, and what results are expected when the job has been completed. The client should be aware of the people moving through the house, the noises the equipment makes both during testing and during the work, and how long the project will take.

The QCI needs to check those records and make sure the fundamental questions have been asked and answered and determine if there is anything missing. Use the opportunity to check with the homeowner for their understanding of the elements of the job.

The other interview step is talking with the crew leader. It's one thing if you are part of the same organization. It will be different if you are an independent QCI. But in either case, the QCI takes the ultimate responsibility for the quality of the job. It is your name and your reputation that is riding on the outcome. You can't just say, "But the crew leader didn't tell me about that!" The questions may seem obvious or redundant or boring, but they need to be asked. Watch the body language and the tone of voice. Listen to what he or she is saying and what he or she is not saying. Never assume! Except for one thing: I think you have to be able to assume that the crew leader and the crew wants to do a good job, the best job they are capable of.

If you know the crew and the crew leader, what you need is to be introduced to the job and the homeowner. You have the paperwork that tells you the basic facts about the house, the work order that tells you what is supposed to be done, and the energy auditor's performance reports. What do you know about the energy auditor? Is he/she familiar and what do you know about his or her capabilities? Do they work for the same organization or are they a part of a different organization, one that might have different rules? But even if you know the crew and the crew leader and the auditor, don't assume that every detail is right especially when it comes down to the health and safety of the job or house.

I remember clearly the first time I worked on a roof, crawling around on my hands and knees and being overly fearful. After a while I got used to it, and walked around easily. But I was still enough of a rookie to be cautious. If you get immune to the dangers and take things for granted, accidents happen.

Task 1.2 Assure Worker Professionalism

What is professionalism? A professional has been trained and developed skill in performing a job or a task to a high level. An amateur can certainly be skilled, but a professional has to be good enough at the job to get paid to perform it. And it is a fact of life, that if a professional does a job badly, he or she may not be doing it very long.

As a QCI, you have to be able to easily recognize when a job has been performed to professional standards, performed so well that the worker is providing top of the line service.

Professionalism takes time, repetition of a task over and over again, until it can be done to perfection, setting up and taking down a blower door, duct tester, or combustion analyzer over and over again until you can do it with your eyes closed. And when you do make a mistake in the process, it will only take a moment to recognize the mistake and correct it.

How will you recognize worker professionalism? You're going to have to tell your proctor during the field-test how you recognize the professionalism of the workers. It will be evident in small things like how they put their tools away, how they maintain the cleanliness on the job site, how they remember to put everything back in the house just the way it was (or better) before they leave. It will be evident in how they fill out and retain their paperwork. There won't be a bunch of coffee stains on the pages, there won't be cigarette butts in the yard, and you're likely to get a comment from the Edna, the homeowner, like, "They were such nice people!"

If there is an evident lack of professionalism on a project, how will you address it? Ensuring Worker Professionalism is the essence of Domain IV.

This task focuses on what are commonly known as "soft" skills like communications through the client interview techniques, observation of the workers and the job site, and remaining tactful when things don't go smoothly. This task requires an awareness of your employer's requirements and code of conduct. It also requires the ability to provide positive reinforcement techniques, providing praise and encouragement when it is deserved.

BPI QCI Field Test

- Candidate noted any damage potential caused by workers;
- Candidate mentioned the need to perform spot checks on in-process work;
- Candidate identified work that does not meet acceptable practices.

1.2.1 Client or occupant interview

Occupants live in and with the building – they are a primary source of information about it. You need to listen carefully to what they say and what they don't say. They may know a lot about the house. They may have had it built or remodeled it from top to bottom or they may have moved in less than a year ago. They may know a lot about construction or they may know nothing. They may care a great deal about how you treat their home or they may not care at all and want you to get this over with and get out!

To do an effective client interview you need to have an effective questionnaire and you need to be familiar enough with it that you can leave it at home and move through the questions naturally like a conversation with an old friend. This interview not only helps you understand some of the issues that you need to move forward, it will also establish your rapport with the homeowner and might turn out to be a solution to future problems. It may be that your company or organization has a client interview form that you need to use. Or you may have to create your own. Homeowners come in all flavors from friendly and helpful to nasty and impatient. You need to use your relationship building skills to make the job go as smoothly as possible.

A client interview can take a significant amount of time and the information collected can be extremely useful so you don't want to rush it. The actual client or occupant interview should be done by the energy auditor prior to the generation of the work order for the house. So the task for the QCI is to review the input of the existing interview documentation and seek elements that need clarification or reinforcement and use those points during your field-testing to demonstrate to the proctor your skill in communicating with the occupant. Since you are going to be talking to the occupant while the project is proceeding, you probably will probe their satisfaction with what has happened so far and if they have any questions or complaints. Is the project proceeding at they expected? Have their been any surprises - positive or negative? Have they been provided with a clear description of is happening, what will be happening, and what they can expect from the final results?

People provide all sorts of clues about the lives and their passions or enthusiasms throughout the house. You definitely don't want to pry or ask questions that the occupant would rather not answer. You don't want to ask her when the baby is due unless you are sure that she is pregnant! You do want to know if the occupant has health issues that are caused by the house. For example, if they wake up every morning with a headache, it may mean that the furnace or water heater back drafts because they close the bedroom doors at night, cutting off combustion air to the appliances.

The kitchen table is a great place for an occupant interview. It is home base in many houses. You can see the pictures on the refrigerator that might give you a clue as to a favorite sports team or grandchildren. Remember too that you are acting like a surgeon regarding the general disorganization of the house. A surgeon only sees the body being worked on. Not the clothes. When you are working on a house, you need to be looking at anything that impacts the energy performance - not the collection of movie posters or the unmade beds. You do need to take into account anything that will impact the work you are going to do - large display cabinets in front of the attic access hatch or a historic, crystal chandelier near where you want to put the blower door!

The point during the field test is to demonstrate that you are skilled at communicating and establishing rapport with the occupant. (Note that even though your discussion with the proctor is one sided, you should be able to demonstrate your communication skills by constantly informing him or her about what you are doing and why you are doing it.)

Sample Interview Questions:

- When was this home built?
- How long have you lived here?
- How many people live here?
- How many pets?
- Have you done any remodeling or structural changes?
- What energy efficiency actions have you taken already?
- If so, have you noticed any changes in comfort?
- Does your house have particularly hot or cold rooms or areas?
- Where and when does this occur?
- What sort of heating and cooling systems do you have? (Hot air, hot water, central AC, window AC, etc.)
- Fuel type?
- How old is the system?
- When was it last serviced/cleaned and tuned?
- How often do you replace your filter (if forced air)?
- Have you had the chimney cleaned?
- Where is (are) the thermostat(s)?
- What type(s)?
- Do you program them?
- Do you use other appliances for comfort heat - space heaters, wood stoves, etc.?
- If so, what type and where are they located?
- If so, how often do you use them? For ambiance, infrequently, regularly, etc.?
- If a wood fired appliance is used regularly, how many cords/season?
- Do you have any unvented gas appliances?
- Do you use other appliances for comfort cooling - windows air conditioners, fans, etc.?
- If so, what type and where are they located?
- If they are windows air conditioners, do you remove or cover them for the winter?

1.2.2 Evaluate the Job Site

The cleanliness and organization of a job site is a sure sign of the professionalism of the workers. You can tell almost instantly how much they care about their work and the organization that they work for if tools are put away, trash picked up, and new materials stacked and protected from the weather. There are great programs to recycle construction waste. (http://www.cdrecycling.org/ or http://www.epa.gov/epawaste/conserve/imr/cdm/) This is not a part of the QCI exam, but it is definitely part of quality control. The less we have to put into the landfills the better off we'll all be. Besides that it can be great for your public relations and profitable.

At the very least, job sites should include a means of collecting waste. Trash cans, trash bags, places to put construction waste - loose pieces of insulation, cut off ends of wood, scrap wires - the workers should not be walking on or balancing their ladders on that stuff. It's unprofessional, messy, and dangerous. There shouldn't be empty - or partially empty - coffee cups perched here and there. There shouldn't be cigarette butts ground into the dirt. What about bathroom facilities? You definitely don't want to be using the homeowners' bathrooms. If there are portable facilities, are they clean and well maintained? The QCI's responsibilities span everything about the project from how the insulation is installed, the air leaks sealed and the porta-potties maintained! The bottom line is: are you proud of the job site?

1.2.3 Workers Familiar with Code of Conduct

Does your organization have a written Code of Conduct? It should if it doesn't. Job site conduct reflects directly on the professionalism of the project. All workers should understand the Code of Conduct and subcontractors should be informed of the policies. This is a sample of Code of Conduct:

- Hard hats are required in all areas depending on the application of safety rules. Hard hats are part of the set of personal protective equipment along with eye protection, CO monitors, gloves, steel toed shoes;
- Clothing and footwear must meet OSHA standards. T-shirts must not display slogans or images that may be offensive to other tradesmen, to the homeowner, or to the community;
- The use of obscene, profane, abusive, or threatening language is not permitted either orally or in writing;
- Workers are expected to conduct themselves in a dignified manner at all times;
- Any behavior that may be viewed as harassment will not be tolerated;
- There is zero tolerance for drugs, alcohol, and any other intoxicants. Possession or use is prohibited and will be grounds for dismissal;
- Smoking is only permitted in designated areas within the job site. If the homeowner has a more restrictive policy about smoking, that policy prevails;
- Loud or disruptive music on the job site is not permitted.

A. Tamasin Sterner of Pure Energy Coach (http://www.pureenergycoach.com/) has an exemplary code of conduct that she makes everyone who works for her sign.

PURE ENERGY
QUALITY ASSURANCE

IV. INTERACTION WITH CUSTOMERS

Pure Energy Coach representatives and contractors shall not engage in inappropriate relationships with customers or denigrate customers in any manner.

To that end, Pure Energy Coach representatives and contractors are prohibited from:

- dating or engaging in personal relationships with customers
- initiating physical contact with customers
- making derogatory comments to or about customers
- telling insulting jokes to customers
- utilizing pejorative slang terms representative of certain groups
- discriminating against customers on the basis of race, color, religion, creed, sex, gender, gender identity, sexual orientation, age, ancestry, nationality, national origin, disability, marital status, veteran status, or any other status protected by applicable federal, state, and local law
- utilizing email, office mail, or voicemail in a negative or derogatory manner
- engaging in hostile acts or acts of aggression, including but not limited to, fighting, kicking, or other acts that could harm another individual or property

<u>ACCEPTABLE ON-SITE BEHAVIOR</u>

- We are there to fully listen to the customer and inspect the work. We are not there to tell the customer what to do.
- We offer alternative ways of using home systems, and we share what we know about the savings associated with the alternative ways vs. the way the customer is currently operating their home systems.
- Stay neutral about religion, politics or any other topic that may raise emotions. We avoid bumper stickers on our vehicles and avoid wearing clothing with printed messages.
- When we represent the utility programs, we stay positive about all team players: the utility, the contractors and ourselves.

PURE ENERGY
QUALITY ASSURANCE

- We don't have the authority to require contractors to return to the job. We let the customer know we will report what they are telling us but don't promise them any particular action. Stay subjective and don't offer opinions.

- When taking photos of the customer's home and items, we ask permission first.

- Respect the customer's time. When you have a firm appointment time and find you will be more than 5 minutes late, call the customer.

- Use your own equipment and tools unless the customer insists you use theirs.

- Never use equipment supplied by utilities on non-utility jobs.

- Our comfort, health, and safety is important. If we don't feel comfortable in a customer's home, we leave as gracefully as we can.

INSPECTION NOTES GUIDANCE

Background

The purpose of the written inspection reports is to provide an explanation to the utilities of problems seen in the field. They are used as an official record of what we observed.

Keep it short and to the point. Report the problems with a job. If it was a straight Pass, we do not need any narrative.

Our comments and recommendations to the contractor are based on building science, energy savings potential, BPI Standards (which is code-based and materials installation specifications based), and program based.

Example: We will avoid commenting on the condition of window unless treating them will reduce energy use for sure.

If the windows are not in an area of the home that is under high pressure, then treating them won't meet Pure Energy Standards. If they are in the high pressure areas of the home, follow this priority: Seal high first, low second, none in between in the neutral pressure plane.

PURE ENERGY
QUALITY ASSURANCE

GUIDANCE FOR MAKING YOUR INSPECTION NOTES DECISIONS

- The contractor should do air sealing in this order: Attic floor air sealing first, (not windows first, not basements first), then attached garages, then basements/crawlspaces. Air sealing should not be done in the neutral pressure plane if the attic floor was left unsealed.

- Living space to attic air leakage can be found:
  - By using the blower door and smoke,
  - By using the blower door and zone pressure tests,
  - Pressurize the building with the blower door and use your thermal imager in the attic,
  - Or by visual inspection, knowing the typical locations where buildings leak air.

Be sure to tell the contractor which of the above methods you used to come to your conclusion that the air barrier is still incomplete.

- The passing zone pressure test is dependent on the air leakage from the zone to the outdoors due to the ratio of pressure to air leakage. So, if the post-work attic zone pressure test is not 50 Pa, there is likely still air leakage between the attic and the rest of the building.

  If you see air leaks, the job fails. If you don't, and you are not doing air leakage testing yourself, you can only comment on the test result. Be sure to follow program rules for a pass/fail criteria for zone pressure tests.

- Air sealing CAZs from the outdoors is only encouraged when
  - attics are fully separated from the living spaces, or
  - when atmospheric draft combustion appliances are in the CAZ, or
  - there is no apparent moisture problem in the basement or crawl.

- When reporting a work quality issue, report what you are sure you saw, not what you think you saw. We are an important part of a team - give the contractors the benefit of the doubt. Follow the Golden Rule: Treat the contractors as you would like to be treated.

- Please keep your notes brief, simple, factual, unemotional and free of opinions. If you are not sure, don't mention it.

- Do not write the word "liability" anywhere.

COMMERCIAL AND ADMINISTRATIVE 51

1.2.4 Observe Worker Behavior

The key to observing worker behavior is to be subtle about it! You don't want to hang over their shoulders or stare at them while they do a job. But as you are moving around the job site, checking on the elements of the work order, talking with the crew leader, you should also be observing how the workers are behaving, doing their jobs, staying within the boundaries of the code of conduct. At the same time, you don't want to make snap judgments. No one is perfect - that's why there is a job for quality control inspectors! But you certainly want the job site behavior to be above average. Quality of craftsmanship, safety, and the effective use of materials are elements to pay particular attention to. Most workers want to do a good job. They want to behave professionally. If they don't, they should be retrained or removed.

Task 1.3 Address Work Problems

There is a section on the field guide entitled Address Work Problems specifically concerned with discrepancies with the work plan, the need to run diagnostic tests, and checking for missing opportunities.

General work problems can certainly occur at any time during a project, and it is unlikely that it will be limited to these three! The Foreman or Crew Leader should handle the majority of job site problems, but he or she may turn to you when there are more difficult issues. The most difficult ones usually involve people, people doing things they shouldn't or not doing things they should. Some of these issues can be solved by simply following the company or program rules that should have been laid out and explained in advance.

Observe sequencing of the components being installed: If cellulose is blown into an attic before the air sealing of the attic floor, that's going to be a problem. If a new furnace is installed prior to the reduction of the energy load of the house, it will be oversized and that's going to be a problem.

Check materials being installed (proper materials and quality of materials): If cellulose has been specified for the wall insulation, and the crew is blowing in fiberglass, that is going to be a problem. If the work order specifies a one half-sone bathroom fan and a three-sone fan is installed instead, that's going to be a problem.

Discuss missed opportunities with the auditor: If the auditor didn't have time to explore the back recesses of the attic which turn out to be uninsulated, that's going to be a problem.

Discuss issues with the crew chief: If the crew chief is not having the crew pull on their booties to protect the floors in the house, that could be a problem. If the crew chief isn't able to get the job done because he/she doesn't have enough crewmembers that could be a problem.

Document process issues and missed opportunities for change orders: Objects that aren't moved out of the way prior to work, fire in a fireplace or wood-stove, combustible items around a furnace or boiler, uninsulated attic hatches, and mis-location of the thermal boundary, etc. will impede the successful completion of the weatherization project.

Work orders may need to be revised and retrofit work redirected, and there is undoubtedly a process for doing that - who has the authority to make the decisions? Ideally those changes can be addressed quickly and efficiently, but it will take some skill on the part of the QCI.

————————————————

BPI QCI Field Test

- Reviewed the work against the work plan;
- Identified need to run diagnostic tests;
- Checked for missed opportunities.

1.3.1 In-Process Visual Inspection

Exterior Walk Around

Like all the rest of this, having a repeatable, standardized process and a checklist makes the visual inspection reliable and useful. There is no single approach for this, so do what works best for you. All of this and more should have been done by the energy auditor when they scoped out the job. The process here is to familiarize yourself with the project and verify that nothing obvious has been missed as well as to verify that work that has been or is being done on the house is being done in a professional manner.

Personally I like to take a picture of the house number on the mailbox or by the front door. I use it as a separator from all the other houses I take pictures of. Then I step back, maybe across the street, so that I can see all of the front of the house and take another picture. That's the one I can use for the cover of a report. At that point I also like to take a compass reading so that I can note which direction the front door faces. And I look at tree shading, wind protection, neighboring houses, roof penetrations, the color of the house, the condition it's in, all the stuff that will tell you a great deal about the house and the occupants.

You will also want to turn on your personal CO detector and note any elevated level so that you can compare it to the level inside the house.

I walk around the house in a counter-clockwise direction. It's just my choice. If you'd prefer to go the other way, by all means do so. Just pick a direction and stick to it. **Do it the same way every time.**

As you walk around the house, note the condition of the windows and doors, the drainage around the foundation, the location of the gas meter or oil filler pipe, note the exhaust vents so you can look for the fans when you get inside, and check out the exhaust vents from the combustion appliances. Is the chimney 2 feet higher than anything in a 10-foot radius (the 10:2 rule)? If it's not it may not be drafting properly.

Check the air conditioning condenser and note its specifications and condition. Are there bushes or other objects that will impact the operation of the condenser and reduce its efficiency? Is there any other equipment outside the house that will impact the conditions inside? What about the dryer vent? Is it clear?

All of these items and more should be included on your checklist. If insulation was blown into the walls, was it done in a professional manner? Were the fill holes carefully repaired? Did the installers clean up after themselves? Are there paint flakes on the ground?

Interior walk-around

When you're ready to enter the house be it through the front, back or side door of the house, make sure you either take your shoes off or put on shoe covers, hopefully following the guide of the occupants or the workers.

Walk around the inside of the house the same way you walked around the outside. When I walk around the inside of the house in a counter-clockwise direction, I am able to always keep my right shoulder parallel to the outside wall. Sometimes there are inconspicuous closets or stairs that would be hidden and missed if I don't follow a consistent route.

While you are walking you should be monitoring your personal CO detector and your gas leak detector even if the house does not use natural gas or propane. An excellent auditor once discovered a detectable level of gas in a house that didn't use natural gas. He alerted the gas company and they discovered a gas leak under the street outside the house. Gas fumes were migrating through air leaks around the utility feeds into the house. If he hadn't detected it, the entire block could have blown sky-high.

1.3.2 Review the Work Against the Work Order

The work order is the guide to the project. You should have a copy in your client file. The Crew Leader should have one in his client file. If R-30 is required for finished attic insulation, you should check to make sure that R-30 insulation was installed in the attic. Let's go back to that question at the beginning where blowing R-30 insulation through the entire attic was specified but there already was R-13 batt insulation installed which the workers left in place. What do you do?

Did the energy auditor miss the existing insulation? You have to find out. If he did, maybe he is not doing a thorough job, which would affect other projects. Maybe he or she needs more training. Perhaps the documenting process is not effective and the auditor simply wrote down, "Increase attic insulation to R-30" which might mean that the crew leader misinterpreted the work order. Was the batt insulation installed properly in the first place with the kraft paper facing the heated space? Were the holes in the attic plane properly sealed? Extra insulation should be a good thing for the homeowner, but not if there are other problems in what was already there. And it will cost the organization more money than it should.

The items on the work order should be listed in the order that they should be performed.

What is the average R value of a 1200 ft^2 ceiling if most of it is insulated to R-30 but the 8 ft^2 attic hatch is only R-1? The Residential Energy Dynamics (http://bit.ly/1ecdCDQ)website has a free tool for figuring this out.

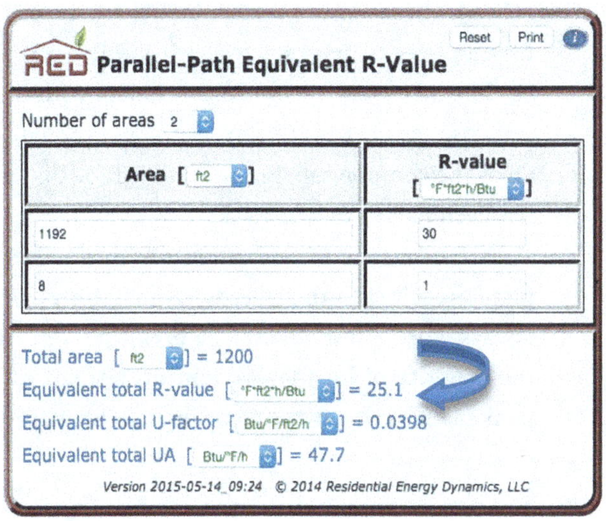

The area of the attic is 1200 less the area of the hatch: 1192. The average R-value of the attic would be only 25.1! Not exactly the over insulated example we were working with, but interesting just the same. This would indicate that the average R-30 requirement of the work order had not been met.

Our example goes the other way. What would be the average R-value if 800 square feet were insulated to R-30 and 400 square feet had the R-30 added to the existing R-13? What would be the average R value?

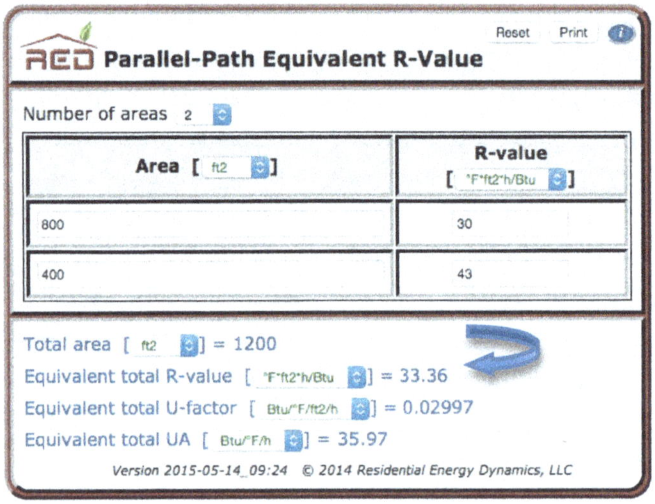

Looks like R-33.36. Not that much above the required average R-value. And if the hatch was still only R-1, the average drops below 30 again. The RED tool will allow you to look at 10 different areas to arrive at an average.

1.3.3 Check Materials

The 'Gofer': The Gofer is the guy that has to be sent out over and over again to get the materials that have been forgotten. Work has to stop because a component is missing. Work has to wait until the Gofer returns with it. The fewer times the work has to stop, the smoother the project will go, the more coherent the work will be, the sooner Edna can get back to her life and enjoy her more comfortable living.

To avoid overuse of the Gofer, plan ahead. Experience will help to generate a checklist. What tools and materials are needed for every job? What tools and materials are needed for today's job? Are the batteries in the test equipment charged? Are there spare batteries? Pilots have checklists that they are trained to review before every single flight they make even if they have flown thousands of times. It is common to assume that you know what should be there. I once worked with a guy that made bathroom fans. There was a quality control checklist for each fan. One day I found him at the copier machine copying the QC checklist and I asked him what he was doing. Every checkbox was checked off! He told me that the fans were new. He'd just made them. There couldn't be anything wrong with them. If you don't look in the truck and see that there actually is a 643M roof vent, how do you know that it's really there? If the crew leader asks for one on the job site, you can't say, "It was here this morning!" The check mark on the checklist should verify that. (Unless you're like my co-worker at the copier!)

There should be two checklists. One is the tools and materials that are needed on every job. The other is the tools and materials that are needed on today's job. As the QCI, you should ask to see both lists, and verify that they are both checked off.

The materials list should coordinate with the job work order. If the auditor calculates that 30 bags of cellulose are needed for a project, there should be 35 on the truck. The materials checklist should include all of the components that are needed for that project and the Crew Leader should check that everything has been loaded on the truck.

Not using checklists is unprofessional. There are some extraordinary people who can maintain complete checklists in their heads. A long time ago I worked with a guy called Cliff that was like that. Cliff always wore a shirt that had a breast pocket. Although it was always stuffed full of little slips of paper, the majority of his files were in his head. Cliff remembered every part of every product that the company made. He was the company's archive of information. The problem was that when he wasn't there, neither was all that information. Job security is great, but it doesn't make for a smooth running or efficient operation. And it doesn't allow for verifiable quality control.

1.3.4 Observe Worker Skills

Knowledge, Skills and Attitude/Ability. Each worker on a project brings special skills. Even the simplest tasks can be done with skill. Sweeping the floor can make it clean or leave it scattered with debris. With that skill comes the knowledge of the difference between the floor of an operating room and the floor of a construction site. The sweeper needs to know. The crew leader needs to know. The quality control inspector needs to know.

Organizations like BPI and RESNET have specified the difference between Grades 1, 2, and 3 installed insulation. The experts got together and came up with the criteria to estimate the quality of the installation. Quite simply Grade 1 is a perfect installation and Grade 3 is barely acceptable. Grade 1 provides the most value from the material. Grade 3 is almost a waste of time.

A bathroom fan that is connected to the light switch in a bathroom so that it runs only when the bathroom is occupied is a Grade 3 installation. Its effectiveness is just about the same as having no fan at all. And if the fan is not ducted all the way to the outside, the QCI should fail the job.

Each skill involved in improving the energy efficiency of a house has a similar scale. And the scales adjust as the project moves forward. It is acceptable to store materials or tools on a job site as long as they meet the criteria of that moment in the job. When the job has been completed, the criteria changes, nothing should be left behind except that materials that went permanently into the job.

1.3.5 Verify the Condition and Capacity of the Equipment

As you are going through the house, you should address the issue of the condition of the space-conditioning equipment such as the furnace and water heater, but you should also consider the condition of the refrigerator, stove, and oven.

A lot of moisture is produced during the combustion process, and the seams in the flue of an atmospherically venting boiler or furnace may show signs of poor drafting. Even before you start doing your CAZ testing, these white streaks will tell you that this appliance is not drafting as it should and the chances of it spilling are high.

In terms of capacity, when the heating and cooling load of the house is reduced by improving the insulation and reducing the air leakage, the load on the heating and cooling equipment is also reduced. It seems awkward, but the ideal operation for traditional heating and cooling equipment is to be running all the time – at the design temperatures. If the design temperature is 6 °F, for example, the furnace will run most efficiently if it is running constantly at that temperature. It won't have to reheat the flue. Ninety-nine percent of the time the temperature won't get colder than the design temperature (based on a 30 year average). Most of the time the furnace or boiler will not be running all the time. If the temperature drops below the design temperature, well . . . the temperature in the house may drop a degree or two below 68 °F (if that is your thermostat setting). So if the thermal load of the house has been reduced and the original furnace or boiler was oversized (which is typical), it will now be even more oversized and will run less and be less efficient.

If you calculate the total U value of the house, you can easily use it to properly size the heating system. You need to total up the conductive UA values of all the surfaces – walls, windows, floors, ceilings, etc. – and add the convective U. The total U value times the design delta T will provide the BTU lost per hour at the design temperature, i.e. the number of BTUs that the heating system will have to supply. If the total U, for example, turned out to be 425 and the design temperature is 6 °F, the delta T is 62. 62 times 425 equals 26,350 BTUs/hour. So if the furnace puts out 110,000 BTUs/hour, it is overwhelmingly oversized.

The problem is even worse for cooling equipment. If it has been oversized, the system simply won't run long enough to dry out the air. The occupant lowers the thermostat setting, which makes the air cold and damp so people have a tendency to add window air conditioners even if they have central air! That's not good for energy efficiency.

Another sizing issue has to do with the space in which the combustion appliances are located. You need to have 50 cubic feet of space for every 1000 BTUs. For example if the CAZ has a 40,000 BTU water heater and 110,000 BTU furnace, it would need (150,000/1000) x 50 or 7,500 ft^3. If the basement has 8-foot ceilings the total floor area would be 940 square feet of space, about 20 feet by 47 feet. It is surprising how often the CAZ is too small. Seeking a solution that doesn't break the thermal or pressure boundary can be a challenge. If the combustion appliances are in a closet in a garage in a cold climate, contractors will sometimes provide a heating vent to keep the pipes from freezing. That heat is going directly to the outside. Finding a supply register in such places can be a challenge when you are trying to do a duct test!

1.3.6 Determine the need to conduct diagnostic tests

If there is mold in the living room, you don't need to analyze what kind of mold it is. It should just be removed. If you can see gaps in the ductwork, you don't need to test the ducting. They should just be sealed. There are times when diagnostic testing is important and times when just getting the job done is what you need to do. The fact is that every house could be a science experiment. We can learn something from every house we work on. But life doesn't often provide that time. If you have to audit two houses or more per day, you just have to get it done.

There are certainly times when diagnostic tests will save you time. If your program has a limit to how tight the house should be, blower door guided air sealing can help you to determine when to stop looking for holes. Zonal pressure diagnostics can help determine the rooms that have the majority of holes. Combustion safety testing should be done every day so that the home occupants will be safe when you're not there working on the house. Most diagnostic tests can be done at the beginning and the end – test in and test out.

1.3.7 Conduct diagnostic tests

You need to know how to do all of the basic building science diagnostic tests even if you don't use them every day. You not only need to know how to do them. You need to know why you are doing them and what the information means. If you don't know what the information means, you won't know when you are doing the test incorrectly, when there is a problem with your tool, or when there is something truly wrong.

Colin Genge, the president of Retrotec, recommends playing with your manometer. Get to know the difference between a hose that is pinched in a door and hose that someone is standing on. Hold the end of the hose over a puddle and see if it sucks up the water. Should it? Hold the end of the hose out of the car window pointing into the wind and point away from the wind. Watch the reading while you wave the hose up and down. Connect a short hose from one Reference tap to the other. Do both channels read the same? What would happen to your blower door reading if you attach the hose from the fan to the Reference tap instead of the Input?

Don't just memorize how to hook up the equipment up. If you don't know why we test houses at 50 Pascals, ask someone! Understanding is the key to doing this job right and there is an amazing amount of stuff to understand.

You need to know these basic tests:

Combustion Safety and Efficiency Tests- gas leakage, flue condition, and system components

- DHW temp
- CAZ Testing - manometer
- Draft based on temperature
- Spillage
- CO
- Delta T rise
- Oven test
- Duct testing
- Blower door testing
- Pressure pan testing
- Zonal pressure to the attic
- Exhaust fan flow

The majority of diagnostic tests should have been performed by the energy auditor. They are, however, included in the QCI field exam so that the candidate can demonstrate that he/she has the knowledge to check the work if necessary. It is a great deal to accomplish so you are going to need to budget your time wisely. The actual steps involved are based on the BPI Building Analyst Standard and not on the BPI 1200 Standard that actually outlines the performance of the tasks in a more detailed way.

Before beginning the tests you should refer to the original energy auditor's report and the work order for the project. You can compare your results with those. As a Quality Control Inspector, you need to demonstrate that you know how to read reports and check quality. You are going to want to be sure that all the equipment have been calibrated, that the batteries are charged, that you are familiar with the particular tools that you are using, and that everything is functional. It would be a good idea to mention to the proctor that you have checked the calibration stickers or screen notices and that everything is up-to-date. This is a quality control test after all.

You are definitely going to want to use a checklist for all of this. There is one included with this book, but you may be more familiar doing the tests in a different order. Because of that, you may want to take the checklist supplied and modify it to meet your system or the system from the agency or company that you work with. Remember that BPI's fundamental rule is, "Do no harm." This is the most important part of any audit. Houses can explode and people can be hurt or worse if the combustion safety testing is not done properly.

You may want to set some things up in the house prior to beginning the field test. It may be worthwhile to unpack the blower door and the duct tester. Get out the manometers and hoses and have them laid out near where you are going to need to use them. Make sure that you have the information on the footprint and volume of the house. Make sure that you know where all the heating supply and return registers are and where the ducting is inside and outside of the pressure boundary.

You need to monitor the ambient CO throughout the test. This is best done with a personal CO monitor that you carry on your person. You are going to need to be sure that it is on when you are outside the building and notify the proctor what you get for a reading. If there is an elevated level of CO outside the building (for example if the house is in an area with a lot of traffic), you can be sure that it is not being generated inside the building. You want to let the proctor know that you understand that.

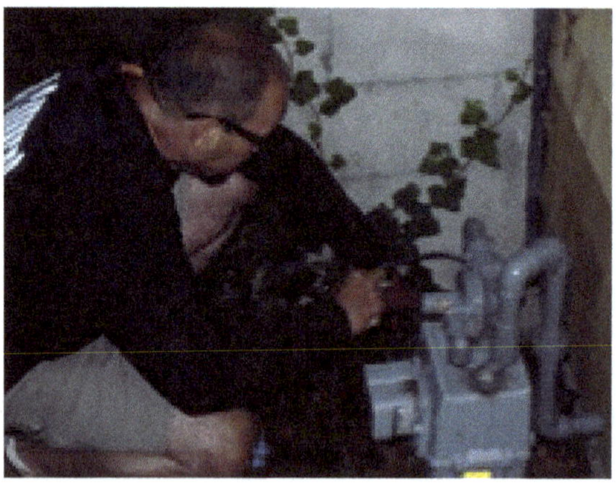

If the combustion appliances are gas or propane fired, you need to check the gas connections outside the house and continue to test them on the inside. Be sure to tell the proctor that you are starting your test at the origination of the gas line coming into the house and following it out to each appliance, testing every fitting. Some combustible gas detectors have two settings; they start out in their most sensitive setting. If the device goes off, shifting to a less sensitive setting provides verification that there might be a problem. At that point, bubble fluid should be used, and if a leak is indicated, a piece of surveyor's tape or something similar should be used to mark the location. With all the testing to be covered by this test, it would be impossible in most houses to test all the connections. Demonstrate to the proctor as quickly as possible that you know the technique, what to do with the results, and hope that he/she lets you move on. The proctor cannot provide any technical answers, so keep going until he/she tells you move on. Keep the gas analyzer operating until you have finished setting up the house. There have been times when there is combustible gas in the house coming through the utility feed holes from the street. You may save a life or a neighborhood!

If you walked around the outside of the house in a clockwise direction, walk around the inside of the house in a clockwise direction, checking all the exterior doors and windows and making sure that they are closed and locked. All the interior doors (including closets) should be open. There are many air leaks in closets.

While you are walking around with the proctor in tow, make sure that you mention the quality of the air seals on the doors and windows, various ways of determining the level of insulation in the walls[2], and point out areas that are being worked on and explain why. Make sure the combustion appliances are turned off, giving the flues a chance to cool and making sure that they won't fire while you are doing the blower door testing.

Blower Door Testing: Set up the blower door and perform your test. Make sure that you PLACE the exterior hose outside the building and away from the airflow from the fan and out of any puddles and pointing down so water can't get into the end of the hose. This can be a single point test, simply measure the flow at 50 Pascals. Or you could connect the system to your computer and perform a multi-point test, taking pressure readings at multiple pressure points for more accuracy. But because of the limited time available for all the testing, you might just mention the multiple point test process to the proctor to show him/her that you are aware of it. Be sure that you take your baseline measurement and write down the result before clearing the baseline number off the manometer screen. (It's not part of the test, but why do we take a baseline reading? If you don't know or don't understand it, ask somebody.) If you write it down it will verify that the baseline measurement was made when you are checking your results later. It is also a good idea to write down the weather conditions - temperature, rain or snow, etc. Note the ring you are using to the make the measurement.[3] Write down (document) the results[4].

---

[2] *Ask the homeowner if they know.*

Crotchet hooks can be inserted in small holes in a closet and pull out a sample of the insulation.

An infrared camera can be used to measure the temperature on both sides of the wall to determine the R value. (See Residential Energy Dynamics http://bit.ly/1wimk5L)

[3] *Taking a Pressure/Pressure measurement allows you to verify your results. Then take a Pressure/Flow reading. Only use the Pressure/Flow@50 reading if you can't get the house up to pressure. You may experience problems when you are demonstrating the Zonal Pressure Readings if you forget the manometer is set to Pressure/Flow@50.*

Zonal Pressure Diagnostic Testing: For these diagnostic tests, it is vital that you know what the pressure is in the house WRT outside. With the manometer set for a *flow @50* reading you may get erroneous information because the pressure in the house WRT outside may not be exactly 50 Pa. With the blower door set up, you will need to provide some basic zonal pressure diagnostics between the house and the attic and between the house and the basement or crawl space. (You may want to use a second manometer for these tests.) There are numerous ways to do these tests. My preference is to consider the zone that I am sampling as 'outside' of the conditioned space, putting the measurement hose on the Reference port on the manometer so that I am measuring the house with reference to (WRT) the zone just as I did when I was measuring the house with reference to outside. With the blower door running and the house at -50 Pascals with reference to outside, if I measure the house WRT the attic and the reading is -50 Pascals, the attic is well connected to outside and well separated from the conditioned space in the house. If you measure the pressure in the room WRT the house and get a reading of -30 Pascals, you can measure the pressure in that room WRT outside, the reading would be -20 Pascals. These would be similar for the basement, crawl space, or any particular room. When I first started doing these pressure diagnostics tests, I had to draw lots of pictures.

Pressure Pan Test: A pressure pan can be used for several tests. With the blower door running, by covering an outlet or light switch with the pressure pan, the space under the pan effectively becomes a miniature room with air leaks just like a full sized room. Make sure that the blower door is still running so that the house is at -50 Pascals WRT outside. With the manometer connected to the pressure pan, leaks in the pressure boundary can be assessed. For example, if the reading from the pressure pan is -50 Pascals WRT the body of the house, the wall cavity would be wide open to the outside! That would require diagnostics to determine the origin of the leak and air sealing to resolve it. Like zonal diagnostics, the pressure pan can be a very effective means of quickly determining where work needs to be done.

The pressure pan can also be used for general ducting diagnostics. If the ducting is outside the pressure boundary, during the test a direct connection to the outside should be provided to amplify the differential, such as opening a basement window. The closer the pressure is in the ducting to -50 Pascals, the greater the leakage in the ducting. Each grille should be measured individually. It is preferable to measure duct leakage with the duct tester rather than with a pressure pan, but if the duct tester is not available, a pressure pan is a rough substitute. In reality, however, most duct systems need air sealing so rough measurement is probably not worth the time.

Duct Testing: (Note more detailed duct testing information follows in Domain 3, page 63) There are two approaches to duct testing with a dedicated duct tester. One is a measurement of Total Duct Leakage - a summation measurement of all the ducts whether they are located inside or outside the pressure boundary/thermal envelope. The other is Leakage to Outside. This is a measurement of just the leaks that are located outside the pressure boundary/thermal envelope. The latter is done with the blower door running. Total Leakage measurement does not require a blower door.

---

[4] *If you know the area and volume of the house, you can calculate the ACH_{50}. If you plug your CFM_{50} reading into the RED Calc software, it will provide the ACH_{50} along with a lot of other information.* http://bit.ly/1LLpgkk BUT be sure to write the information down.

For either test, all the grilles and registers except one must be sealed or covered so that the fan can pressurize or depressurize all the ducting. It is common practice to use a large return to attach the duct tester ducting. If that is not a reasonable approach, the duct tester ducting can be attached directly to the air handler. The blower door must not be running for the Total Leakage Test. If there is a section of ducting running through a basement or crawl space area, an opening should be provided to directly connect that space to the outside.

There is also some discussion about whether the ducting should be pressurized or depressurized for these tests. Either way works. I have found that depressurizing is less likely to blow the covering or tape off the grilles and registers. And by depressurizing the ducts, the blower door won't have to be turned around because it will be depressurizing the house.

Duct testing is done at 25 Pascals, which is equivalent to approximately 0.1 iwg, which is about the same pressure that the air handler blower puts on the ducting when it is running. That means that leaks discovered by the duct tester will be approximately the same as leaks generated when the air handler is running.

Task 1.4 Evaluate Client Satisfaction

This is one of the most difficult areas to handle. Evaluating Client satisfaction is truly a "soft skill". It is certainly not clinically simple like a high blower door number or unacceptably low pressure in a CAZ! There aren't instruments to measure the real satisfaction of a homeowner. You can only listen to what they tell you. They are relating their feelings! Sometimes this can be simple and sometimes not.

And you are the measuring device! You have emotions too, and you have to filter those feelings out. On top of that, when you are taking a field test, the proctor is evaluating your approach to the homeowner and your ability to understand what they are telling you and how you ask your questions and how you solve their problems.

There is another element from the QCI point of view: if you are an employee of the company performing the work, you have to tread the line carefully between agreeing with complaints from the homeowner and supporting your company and its crew.

Let's go back to that problem posed at the beginning about Edna, the homeowner, complaining about the crew tracking dirt through her house. Do you inform her that she will need to expect a certain amount of dirt when there is construction work? Do you check with the crew leader to see if this is a recurring problem for this crew and recommend training? Do you write up the crew with a disciplinary report? Or do you clean up the area?

I was in a house yesterday that had been beautifully weatherized. The homeowner was happy, but the first thing she mentioned was an area in the basement where the wall had been stained. The basement was unfinished. The only reason to go there was because that's where the washer and dryer were located. But in the crew's enthusiasm to spread mastic on the duct joints they had spilled some that had stained both the wall and the floor. It looked as though they had made an effort to clean up the floor, but mastic is difficult to remove. (That's sort of the point of mastic!) With all the work that had been done effectively and with professionalism, that was a problem for her. Don't think that just because it's not a problem for you that it won't be a problem for the homeowner.

Although it may be true that construction will disturb the house and cause dust, it's certainly not going to satisfy the homeowner if you tell her that she should be more tolerant of workers making a mess! Every construction job does that. It should not be a surprise, and it should be planned for if the homeowner is alerted in advance. There are plenty of tools and equipment to minimize the impact on the house. If you can see that those steps have been taken . . . that there are runners protecting the rugs, shoe coverings are being used, drop cloths are covering furniture, etc. . . . well, maybe the homeowner is exaggerating for some reason or another.

That doesn't mean that you shouldn't check with the crew leader to get another perspective on the problem. Maybe the crew was sloppy setting things up or maybe they neglect wearing their shoe covers because it is too much of an effort to put them on and take them off. Or maybe there was one morning when someone was in a hurry and just forgot.

During the field test there may be a real homeowner or there may be a homeowner stand-in, but you're going to need to talk to them about their satisfaction with the job. The proctor is going to have to evaluate if you "accurately" determined their satisfaction from this conversation. People are rarely, measurably accurate! They're not like blower door tests with a passing or failing ACH_{50} number. They may be reluctant to clearly express their feelings about the project. But if he/she tells you that the project "Sucks!" and you relate to the proctor that you think the homeowner thinks it is going well, that's an obvious disconnect. You can ask the homeowner to describe their satisfaction with the project on a scale of one to ten, ten being fully satisfied. At least that would give you a scalable place to start. But establishing a clear line of communication is the only way to know if someone is really telling you how they feel.

The evaluation of the client satisfaction comes up twice during the BPI field test. The second time it is looking for both the level of satisfaction and the level of comfort.

BPI QCI Field Test

- Asked the homeowner about the work when it was in-process;
- Made note of homeowner's answers;
- Accurately determined the satisfaction level of the homeowner based on the conversation;
- Interviewed the homeowner regarding the level of comfort and satisfaction with the work.

1.4.1 Homeowner Interaction

The QCI field test that replicates what goes on in real life can't take place in a training organization's laboratory! There would have to be a slew of actors participating as the homeowner, the crew leader, the workers, and Ralph, the nosy neighborhood who wants to know what's going on next door! In either case - real or simulated - there is a section of the field exam entitled "Evaluate client satisfaction regarding the in-process work" that includes:

- Candidate asked the homeowner about the work when it was in-process
- Candidate made note of the homeowner's answers
- Candidate accurately determined the satisfaction level of the homeowner based on the conversation

If it is in the laboratory, you will need to at least mention the fact that you would talk with the homeowner during your on-site inspections and you would use that information to make a punch list of their issues. You would mention that in order to ensure the homeowner's overall satisfaction that you would make sure that at the end of the job you would check back with them and perhaps show them that not only were those issues addressed, but many others as well.

The JTA goes beyond these three issues to a more in-depth homeowner interview including noting "homeowner behavioral habits". You certainly shouldn't become a stalker! A female judge occupied one of the houses I used for training. I was working with a great class of students who came from and lived in some pretty tough neighborhoods. (We used to joke about how to evaluate the infiltration through a bullet hole!) During the training we had to poke into every nook and cranny of the house. Her son was showing us around the house, and I asked the judge if she used the open fireplace and she said, "Oh no! I keep it closed off." Her son turned to me and commented, "I will when she dies!"

So the present occupant's behavior was to make sure that the air leaks up the chimney were adequately sealed off . . . but only in a temporary way so that the chimney was not permanently damaged.

Homeowner satisfaction is important. You want them to be so satisfied that they will spread the word that the program is great. Word of mouth advertising is the best you can get. You also want them to be satisfied because their comfort and satisfaction is the point of the program. Making them uncomfortable during the project by stalking them, asking about issues that have nothing to do with the energy efficiency outcome, handling personal objects, or doing anything that intrudes on their personal space and life is not going to result in a comfortable, satisfactory final outcome. The line is thin and the balance difficult, but that's why you're a professional.

Domain II: Quality Checks

Task II.1 Review Client File & Work Scope

Every program will have its own record keeping system. What goes into a client file will vary. There are a variety of software programs to help manage client and project information.

At the very least it should have:

- Client name
- Client street address
- Client phone number
- Client email address

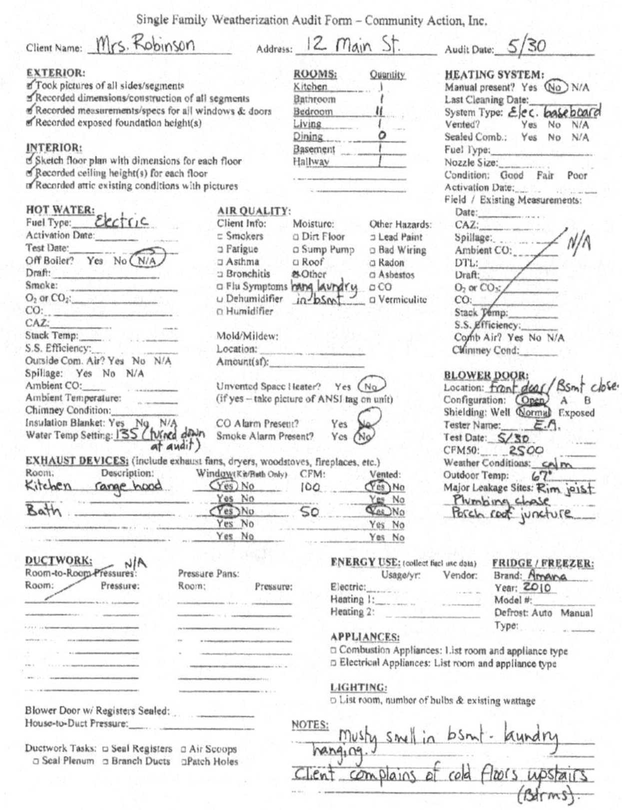

Single Family Weatherization Audit Form – Community Action, Inc.

Client Name: Mrs. Robinson Address: 12 Main St. Audit Date: 5/30

EXTERIOR:
- ☑ Took pictures of all sides/segments
- ☑ Recorded dimensions/construction of all segments
- ☑ Recorded measurements/specs for all windows & doors
- ☑ Recorded exposed foundation height(s)

INTERIOR:
- ☑ Sketch floor plan with dimensions for each floor
- ☑ Recorded ceiling height(s) for each floor
- ☑ Recorded attic existing conditions with pictures

ROOMS:	Quantity
Kitchen	1
Bathroom	1
Bedroom	11
Living	1
Dining	0
Basement	1
Hallway	1

HEATING SYSTEM:
Manual present? Yes (No) N/A
Last Cleaning Date:
System Type: Elec. baseboard
Vented? Yes No N/A
Sealed Comb.: Yes No N/A
Fuel Type:
Nozzle Size:
Condition: Good Fair Poor
Activation Date:
Field / Existing Measurements: Date: CAZ: Spillage: Ambient CO: DTL: Draft: O_2 or CO_2: CO: Stack Temp: S.S. Efficiency: Comb Air? Yes No N/A Chimney Cond: — N/A

HOT WATER:
Fuel Type: Electric
Activation Date:
Test Date:
Off Boiler? Yes No (N/A)
Draft:
Smoke:
O_2 or CO_2:
CO:
CAZ:
Stack Temp:
S.S. Efficiency:
Outside Com. Air? Yes No N/A
Spillage: Yes No N/A
Ambient CO:
Ambient Temperature:
Chimney Condition:
Insulation Blanket: Yes No N/A
Water Temp Setting: 135 (turned down at audit)

AIR QUALITY:

Client Info:	Moisture:	Other Hazards:
☐ Smokers	☐ Dirt Floor	☐ Lead Paint
☐ Fatigue	☐ Sump Pump	☐ Bad Wiring
☐ Asthma	☐ Roof	☐ Radon
☐ Bronchitis	☑ Other	☐ Asbestos
☐ Flu Symptoms	hang laundry	☐ CO
☐ Dehumidifier	in bsmt	☐ Vermiculite
☐ Humidifier		

Mold/Mildew:
Location:
Amount(sf):

Unvented Space Heater? Yes (No)
(if yes – take picture of ANSI tag on unit)

CO Alarm Present? Yes (No)
Smoke Alarm Present? Yes (No)

BLOWER DOOR:
Location: front door / Bsmt close.
Configuration: (Open) A B
Shielding: Well (Normal) Exposed
Tester Name: E.A.
Test Date: 5/30
CFM50: 2500
Weather Conditions: calm
Outdoor Temp: 67°
Major Leakage Sites: Rim joist, Plumbing chase, Porch roof juncture

EXHAUST DEVICES: (include exhaust fans, dryers, woodstoves, fireplaces, etc.)

Room:	Description:	Window (Kit/Bath Only)	CFM:	Vented:
Kitchen	range hood	(Yes) No	100	(Yes) No
		Yes No		Yes No
Bath		(Yes) No	50	(Yes) No
		Yes No		Yes No
		Yes No		Yes No

DUCTWORK: N/A
Room-to-Room Pressures: Room: Pressure:
Pressure Pans: Room: Pressure:

Blower Door w/ Registers Sealed:
House-to-Duct Pressure:

Ductwork Tasks: ☐ Seal Registers ☐ Air Scoops ☐ Seal Plenum ☐ Branch Ducts ☐ Patch Holes

ENERGY USE: (collect fuel use data) Usage/yr: Vendor:
Electric:
Heating 1:
Heating 2:

FRIDGE / FREEZER:
Brand: Amana
Year: 2010
Model #:
Defrost: Auto Manual
Type:

APPLIANCES:
- ☐ Combustion Appliances: List room and appliance type
- ☐ Electrical Appliances: List room and appliance type

LIGHTING:
- ☐ List room, number of bulbs & existing wattage

NOTES: Musty smell in bsmt - laundry hanging. Client complains of cold floors upstairs (Bdrms).

The client file should have as much information about the occupants and the house as can be collected over the phone.

- Year the house was built
- Number of years the occupants have been in the house;
- Number of family members or occupants of the house;
- Main areas of occupants concerns.

Is it a single family framed house? Is there an attached garage?

Let's look at this typical Weatherization Audit Form. This one covers all the necessary bases. Some of these forms are much more elaborate. Remember that you are looking at this as the QCI not the crew leader.

Name, address, and date of the audit are a great place to start. Hopefully the name and the address coordinate with where you are standing and who lives there! And since the form indicates that pictures were taken of all sides of the house those pictures should be in the client file. You should know what the house looks like.

There is a sketch of the house with the dimensions. Note the checklists on this form. Because the little check boxes have been checked, the work was done and the information should be in the file. Look through the file and make sure.

Look down the list of rooms: one kitchen, one bathroom, eleven bedrooms, one living room, no dining room, one basement, and one hallway. Eleven bedrooms? In a one thousand square foot house? Looks like the auditor was using the list of rooms as a checklist, ticking off each room after walking around the house. That would give you some insight into how the auditor works.

Take a look at the definition of the heating system. It's electric in this house. It's great that there is a spot on the form to check off if there is a system manual. Houses should have operating manuals with documentation of all the systems and a logbook of all the changes! That would make checking things and fixing things a whole lot easier.

If the system used fossil fuels, there is a place to note when the appliance was last cleaned (and hopefully tuned).

This is a good, basic list of heating system conditions. You should have a good picture of what you are going to see when you find the system. It would be an improvement if there were a slot to indicate where the system is.

There are some places to note the results of the field-testing of both the heating system and the hot water (DHW) systems. With electric systems in this house, there's nothing to test. If these were fossil fuel systems, all the slots should be filled out unless there is a reasonable exception that must be documented. At a minimum, the information that should be recorded here is the information that your program or organization requires.

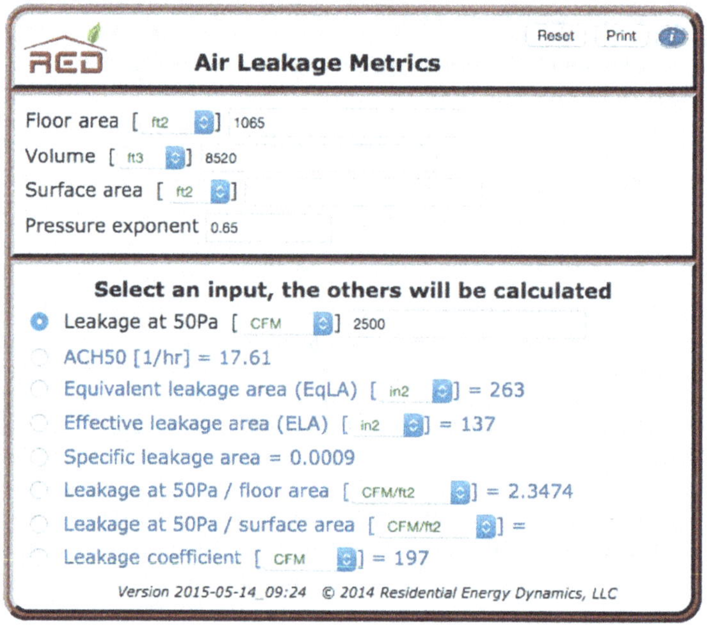

The Air Quality section includes information about the clients, the indications of moisture, and a list of other potential hazards. If they mention flu symptoms, it could be an indication of problems with the flue! Or it could be an indication of carbon monoxide poisoning. It would be an improvement if there were a check box for headaches, another indication of potential CO poisoning.

It's great to have the blower door information available as well. This is a very leaky house. If the volume of the house is approximately 8,520 cubic feet (8 foot ceilings), then the ACH_{50} would be: (2500 x 60)/8520 = 17.6 ACH_{50}. If you divide that by 20 (an approximate n factor), $ACH_{natural}$ is about 0.88. That means that more than 3/4 of the air in the house changes naturally every hour. An acceptable $ACH_{natural}$ would be closer to 0.35.

If we plug the floor area of 1065 square feet and the blower door number of 8520 into the Air Leakage Metrics tool on the RED site, it calculates that the EqLA (Equivalent Leakage Area) is equal to 263 square inches. A 16 inch diameter pizza has an area of about 200 square inches. So if you pushed all the holes in this house together it would be an opening larger than a 16-inch pizza! And that's open all the time. This can be a pretty effective sales tool. A pizza is a much more understandable image than describing it as a 263 square inch hole.

Doing the math helps you to check the auditor's work and the work plan that was developed from the information that he or she collected. If there are inaccurate numbers at the beginning of the project, how can the improvements be proved or justified accurately? The process for making the CFM_{50} measurement must be standardized as well. This form does not indicate whether the measurement was made using the PR/Flow, PR/PR, or PR/Flow@50 setting. It has no indication of whether or not the Baseline was established or what it was. If the auditor does two or three houses a day, he probably won't remember specifically how this particular house was done. He/she will assume that it was done the way all houses are done. In terms of the blower door measurement, the baseline measurement won't make a lot of difference - a few Pascals here or there. But it reflects on the quality of the job.

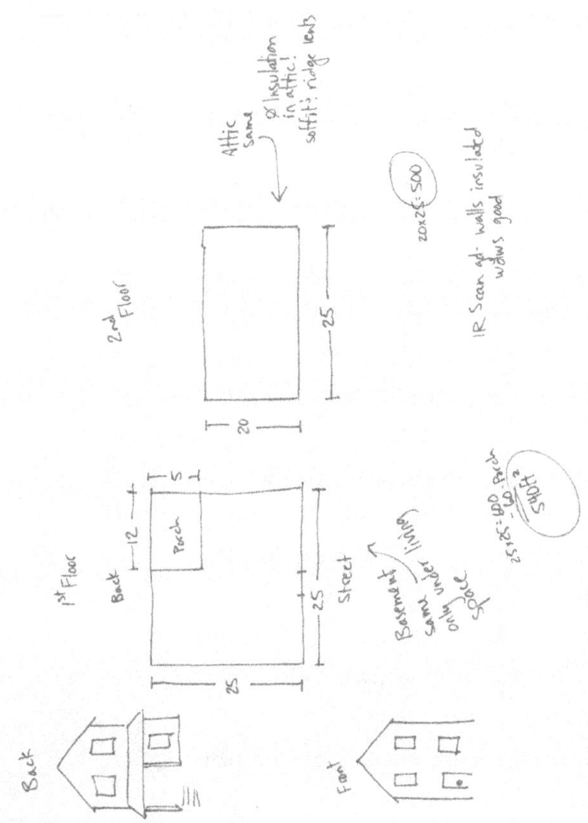

This form has a place for information about the exhaust devices. The form asks for information about exhaust fans, dryers, wood stoves, and fireplaces, etc. These are all items that, when running, might impact the operation of atmospherically vented combustion appliances. It says that there is a 100 cfm fan in the kitchen. Was it measured? How was the 50 cfm fan in the bathroom measured? Round numbers indicate that an assumption was probably made. "I've seen this fan a hundred times before! It's rated at 50 cfm. Why would it be anything else?"

Why was this section included on this checklist? If it is there just to indicate that there are exhaust appliances that might impact the combustion performance, it could just be a checklist with a 'Vented to Outside' column and maybe something about the condition or whether or not it actually runs.

More and more programs are using the ASHRAE 62.2 requirements. To meet that program, the flows of the ventilation systems do have to be measured. Note that the Building Analyst Standard that the QCI exam is based on includes the BAS (Building Airflow Standard) calculation for ventilation. Although this calculation is sadly out of date, many programs still rely on it for determining whether or not mechanical ventilation is required as well sizing the airflow requirement. So it is a necessary to know how to do the calculation.

There is a place on this form for the Energy Use numbers, which is blank. Maybe the homeowner doesn't keep those records. Maybe they don't know where they are. Maybe the auditor didn't get that information from the utility. But since the auditor crossed off the other areas of this form that were not applicable, I would want to know why the utility information is missing.

Take a look at the math calculations on the plan sketches. Are they right? Seems like the auditor missed twenty-five square feet somewhere. Was that on purpose or a mistake? Leaving out 25 square feet leaves out 200 cubic feet with 8-foot high ceilings. Why?

BPI QCI Field Test

- Reviewed work scope (work plan)
- Need for diagnostic tests
- Missed opportunities

Task II.2 Perform Exterior and Interior Inspection

Early on in the project the QCI can review what the Energy Auditor has included on the work order. It is better to find discrepancies early on rather than waiting until the final inspection. It is almost like performing a secondary audit. What will it take to get the attic insulation to average R-49? Will that include the attic hatch? Have the thermal and pressure boundaries been defined properly? How was the insulation in the walls determined? What sorts of combustion appliances are there? How many CAZs?

What did they miss on this house? The snow is evenly spread over the roof of the garage so the interior of the garage is cold. What did they miss in the attic? What can you expect when you walk inside? A snowfall like this is as good as an infrared scan.

When you are walking around the outside of the house, what are you looking for? Always keep in mind your parallel to the Star Trek Primary Directive: Health, Safety, Comfort, and Energy Efficiency. The point of this exercise is to make the house you are working on more energy efficient while at least maintaining, if not improving, the health and safety of the occupants. All of your decisions should be based on that. If Ms Edna, the homeowner, is too frail to remove a window air conditioner, should you do it for her? In real life, you probably would. But in the cold world of certification testing, you need to say that either the homeowner should move it or they should have someone not associated with the program move it for her. From a health and safety viewpoint if you move the air conditioner you might hurt yourself. You might drop it out of the window. You might spill the water inside it on the rug! If Edna can't move it, perhaps she never moves it, and so the house should be tested with it in place. Or maybe you will just need to cover it for your testing. The decisions, what you might consider to be simple common sense on the job, might be different than the cold, clinical decisions on the test.

Make a note of any vents or vent hoods (termination fittings) from fans or dryers. (I found a dollar bill stuck into the outside dryer vent; that told me something about the filter in the dryer!) Do the flaps open? Do they have bird nests built into them? Are they broken and need to be replaced? (This one was mounted sideways.)

Since you are making this tour after the energy auditor and while the crew is working on the house, you are verifying what the audit said needs to be done and the quality of the workmanship that is being done.

Reviewing the work order:

- Has the work listed on the work order been done (or is being done);
- Has the work been performed professionally, work you and your organization can be proud of;
- Has anything been missed that should have been on the work order;
- Note anything else that may seem to have been disrupted or damaged by the process.

You want to be sure that your inspection tour is not disrupting the delicate balance between the work being done and the homeowner's level of comfort.

Task II.3 Evaluate Client Satisfaction

You probably don't know the homeowner personally so you probably don't know the expressions on their face or in their body language when they are happy or satisfied. But life experience should tell you something about their peace of mind. If Edna is yelling at you, you can quickly recognize that she is unhappy. But you can also observe the way she crosses her arms or her facial expression and sense that everything might not be all "skittles and beer" as the British say. (There is an interesting book on body language from an ex-FBI agent, Joe Navarro, called What Every Body is Saying. But if she makes you cookies chats happily about her cats and doesn't complain about the mess and the noise and the nicks in the paint, then you can probably assume that she is reasonably satisfied with what is going on. But the key here is to ***Listen*** to what she says. Let her tell you how she feels.

The best time to talk to the homeowner about the project is when it is complete. That's when you can ask Edna if she thinks it is complete and how she feels about what has just happened in her home. You could do the same thing on your field test. You could ask the stand-in for the homeowner how he or she feels about the project! The proctor can't answer anything technical, but if he/she is standing in for the homeowner, he/she should be able to answer if in his or her opinion the job has been completed to their satisfaction. (Hmm. Interesting quandary for a proctor!)

Task II.4 Determine the Pass/Fail of the Work

While you are walking around the outside and the inside of the house, talking with the homeowner or client, checking the results of the diagnostic tests, you will undoubtedly be making decisions about all that information. That is, in fact, the point. You wouldn't be doing all this watching and testing and questioning if you weren't going to do something with the information! Some of this information will be actionable and some won't. A single, empty drink cup on a windowsill is not something you are likely to want to stop the job for. But insulating the wrong wall because the thermal envelope was not properly defined might be. You are going to have to decide where the line is between Ignore and React.

I did blower door and duct leakage testing on a house not long ago. Neither the house nor the ducts were tight enough to pass the code requirements. The builder couldn't get his certificate of occupancy until the house passed. Sometimes that holds up a move-in date. In this case, it was more than that: the homeowner was dying of cancer. The code said: this is the required performance criterion. My testing said it failed. The builder and the HVAC contractor needed to fix it. The building inspector was standing by the rules. If the house were only going to be occupied for the few months that the original owner was going to live, it wouldn't have made much difference. But houses last for hundreds of years, and the codes and standards are based on the long-term durability and performance of the building.

Fundamentally issues that you can measure are more likely to have clear pass/fail thresholds. If the bathroom fan is required to move 50 cfm and it moves 49, that's a fail. What if you open the bathroom door and the flow increases to 55 cfm? When does common sense come into the pass/fail decision? Houses don't lend themselves to absolute precision. So you do have to blend your knowledge with your ability to reason and the situation. People exercising their own best knowledge, common sense, and the ability to reason have hopefully developed the rules and the standards. (Although at times it seems as though the phrase "A camel is a horse designed by a committee" is particularly true!)

BPI QCI Field Test

- Reviewed results of visual/sensory inspection
- Reviewed results of diagnostic tests
- Compared initial diagnostic results with post-work data
- Determined pass/fail conditions based on the results

Domain III: Conduct Quality Checks Post-work Diagnostic Inspection

Domain III requires the knowledge of a great deal of testing. It would be a rare situation when there would be enough time to perform all of the testing included. You should be able to perform combustion tests on heating systems, domestic water heaters, ovens, stoves, fireplaces, etc. You should be able to perform ventilation system testing and conduct moisture evaluations. And you should be able to conduct electrical safety tests. You should have knowledge of codes and standards adopted by the local jurisdiction, combustion safety protocols, first aid, HVAC systems, moisture and safety issues. And if you are going to make sense of all this, you need to have skill in analytical thinking!

Task III.1 Conduct Health & Safety Tests

Natural gas and propane systems can be dangerous. Installed and working properly, they can provide safe comfort and convenience. They should be checked, monitored and maintained. Like all systems, they are subject to the second Law of Thermodynamics in that they age and tend toward chaos! Natural gas lines are fed to the house from a central system and are connected via a gas meter that monitors the use. Propane is supplied to a tank that is local to the house.

Natural gas pressure in a neighborhood pipeline is 7" wc or 1,743 Pascals. At the customer's gas meter the incoming natural gas passes through another regulator to reduce its pressure to under ¼ pound per square inch (0.25 psi) if this is necessary. (Some services lines carry gas that is already at very low pressure.) This is the normal pressure for natural gas within a household piping system. When an appliance such as a natural-gas furnace or stove is turned on, the gas pressure is slightly higher than the air pressure. A chemical called mercaptan that smells like rotten eggs is added by the gas company. That odor is sure sign of a gas leak.

To test for combustible gas leakage, a Combustible Gas Detector (CGD) is required. It should be classified to *UL 913* Intrinsically *Safe* Apparatus *and Associated* Apparatus *for Use in Class I, II, and III,* Division *1, Hazardous (*Classified*)* Locations or equivalent. It should have a variable click rate or changing tone based on gas concentration levels. (The tick rate provides indication of concentration but only accounts for relative concentration changes, not necessarily identifying hazardous concentration thresholds. It should be capable of providing a digital display of percentage of Lower Explosive Limit (LEL) and/or provide an alarm when detecting combustible gas concentrations exceeding 10% Lower Explosive Limit (LEL).

The problem is that most of the CGD devices sense other things besides combustible gas. They can be activated by pipe dope, cobwebs, or even gases emitted by cooking pasta sauce! Most gas piping in houses have small leaks that may come and go depending on pressures in the system or the temperature of the surrounding air. Gas companies have more accurate and better-calibrated equipment than most energy auditors or QCIs can afford. And if the gas is shut off to the house there will be no heat, hot water, or cooking. On the other hand, if there is a gas leak and the QCI ignores it, it may blow the house up!

So the first thing if you find a gas leak is to be as certain as you can that it is truly a gas leak and not pasta sauce. If the CGD device has a digital display, the LEL or Lower Explosive Limit can be displayed. Other CGD devices have a "locator" ability, increasing the response as the device gets closer to the leak. And some CGD devices have multiple scales. If a gas leak is sensed on the most sensitive scale, shift the setting to the less sensitive setting. If a gas leak is indicated on the less sensitive setting, use some gas leak detection fluid to paint the connection and look for bubbling. Some gas leak solutions have a better tenacity for hanging onto a connection rather than dripping off onto the floor allowing more time for bubbling to be indicated.

If you see burned out matches on top of the water heater, it is a sure sign that it needs to be relit regularly. The first challenge is to determine why. When the occupants can't get hot water, they understand that it probably is the hot water heater and it probably isn't working. So they take off the plate at the bottom and figure out how to relight it. Then they put the plate back because they are hoping that it is a one-time thing, some sort of fluke or aberration or the alignment of the stars. After that happens a couple of times, they make their life simpler and leave the plate off. When flame roll out occurs, one line of defense has been removed and disaster is more likely to occur. The homeowner should be notified of the danger and repair requirement.

This particular water heater was in a closet in the attached garage. Down at the bottom is a warm air supply vent that the HVAC contractor installed to keep the pipes from freezing. It is a very small closet. Remember you need 50 cubic feet per 1000 BTUs. This water heater was not drafting properly and the homeowners couldn't keep it operating properly.

This is outside the building. This was the flue from a water heater in a laundry room. There is a lot of corrosive liquid in flue gases. Because this wasn't drafting properly, the corrosive liquids were just eating up the wall. The conditions on the other side of the wall inside the laundry room were bad as well. There were four clothes dryers all sucking the air out of the room and sucking the flue gases down the chimney. There was an apartment located immediately above the room. There were lots of air passages between the two spaces. The spilled CO in the laundry room was bound to be forced up into the apartment, endangering the people living there.

—————————————

BPI QCI Field Test

- Properly conducted combustion gas leakage testing
- Properly recommended soapy solution to verify positives
- Completed visual inspection of the flue for problems
- Existing heating/cooling components safety concerns
- Measured temperature of the domestic hot water (dhw) at the tap

---

Task III.2 Conduct Diagnostic Tests

You are unlikely to need or have the time to do all the diagnostic tests during a quality control inspection, but you certainly need to know HOW to do them all. Some of these tests you may be doing every day and some you may have only done once or twice before. Before you take the QCI exams make sure that you get your hands on the equipment. Set it up. Take it down. Set it up. Take it down. Understand why you are hooking up the hoses and the sensors the way you are setting them up. Memorizing it is great, but UNDERSTANDING is much better.

———————————————

BPI QCI Field Test

Ability to:

- Perform blower door tests (Great blower door enhancement documentation http://bit.ly/1PiUKUh)
- Perform pressure pan test

- Conduct zone tests
- Perform fan flow tests
- Perform infrared scans
- Perform duct leakage tests
- Perform electric circuit tests
- Perform refrigerant tests[5]
- Conduct appliance tests
- Conduct domestic water heater temperature tests
- Record the results of all tests

Combustion Safety Testing

Set up home in natural conditions

"Natural conditions" means that the home is "at rest". All the exterior windows are closed and locked to get the tightest seal. Make a mental note of the window status before you go around adjusting them. Ideally you will want to put the house back into the same or better condition when you are finished. Exterior doors should also be closed. You will want to alert the homeowner and the crew about that so that they don't interrupt your testing.

Should the interior doors be open? This is an area of some debate amongst building science folks. I believe that the test should measure all the spaces within the thermal envelope and pressure boundary. That can't be done unless all the spaces are connected, i.e. with all the interior doors open (including closets). Starting out with everything open, a baseline measurement can be made before readjusting the doors during the worst case testing. But according to the written Standards, there is no right way or wrong way to do this. So you need to do whatever your program prescribes.

What does the BA Standard say? It doesn't say anything about the interior door settings. *Start with all exterior doors, windows, and fireplace damper(s) closed. Set all combustion appliances to the pilot setting or turn off the service disconnect, including: boiler, furnace, space heaters, and water heater.*

What does the 1200 Standard say? *Close all building exterior doors and windows. Close all CAZ doors. Close the interior doors of all rooms except for rooms with an exhaust fan and rooms with a central forced air system return.*

All the mechanical equipment in the house that will affect the pressures and conditioning of the house must be turned off. This includes bathroom fans, range hoods, heating system, water heater, etc. The flues need to have a chance to cool. Leave your car keys on the water heater so you can't drive away without making sure it has been turned back on.

Proper manometer setup

The goal is to evaluate whether or not the combustion gases are going to be able to flow out of the building under the worst possible conditions, the hardest conditions for the natural and desirable process. High pressure goes to low pressure. The higher the pressure in the CAZ with respect to (WRT) the

[5] To test for refrigerants in an AC system you need to have an EPA 608 license. As a QCI you just need to be familiar with the process and why and when it would be done.

outside the better. To evaluate that situation, the manometer needs to be set up with a hose connected to the outside and since you need to know what the pressure is relative to or in reference to outside, one end of the hose must be outside and the other end should be connected to a Reference tap on the manometer. Just remember what it is that you are trying to find out. It's not magic! Use Channel A of a dual channel manometer for this. Channel A is great for the simple task of just measuring the pressure difference. (Channel B usually has lots of other capabilities and the settings can get confusing.)

1200 Standard: *Using a calibrated manometer or similar pressure measuring device intended for this purpose, measure and record the baseline pressure in the CAZ with reference to (WRT) outside. Compare this measurement with subsequent pressure measurements to determine the greatest negative pressure achievable in the CAZ.*

BPI QCI Field Test

- Measured baseline pressure differential
- Set up home in worst case condition
- All exhaust appliances running
- Correct door adjustments (measured quantitatively or qualitatively)
- Air handler operation impact checked
- Correctly measured (recorded) worst case CAZ depressurization
- Calculated minimum draft pressure based on existing weather conditions
- Check worst case spillage in heating system
- Check worst case spillage in DHW
- Identify time limits for spillage based on BPI Standards
- Determine if appliance passes the spillage test
- Identify what steps should be taken if it doesn't pass
- Measure worst case draft on heating system
- Measure worst case draft on DHW
- Perform necessary tests under natural conditions
- Make appropriate recommendations for work scope based on BPI Standards
- Compare diagnostic results to appropriate table in BPI Standards
- Identify need for further evaluation when other combustion sources exist

Free Combustion Safety Cell Phone App. (http://bit.ly/1FmVFsQ)
Free Building Analyst Cell Phone App. (http://bit.ly/1FmVJca)

Blower Door Testing

Question: Blower door performance is reported on a project but the results are doubtful. How can you calculate the actual performance based on the cfm and the ring used?

The manometer calculates the airflow because it knows the pressure difference through the fan and size of the opening. When you adjust the "Config" or Range setting, you are telling the manometer the size of the hole. So you can calculate the airflow through a hole as long as you know the size of the opening and the pressure difference across it. So you need to know the pressure on Channel B.

Remember that the digital manometers that we use to measure pressures in houses are dual channel. The two channels don't talk to each other unless you physically connect them together. If you set the manometer to PR/PR, you can use the pressure reading on channel B to calculate the airflow. (The manometer does that for you if you set it to PR/FL.) You could plug the Channel B pressure into the following table (Note that the values apply to the TEC Model 3 Blower Door):

Fan Configuration	Calibration Parameters
Open Fan	Flow (cfm) = 506.8 x (Fan Pressure in Pa)$^{.4879}$
Ring A Installed	Flow (cfm) = 190.1 x (Fan Pressure in Pa)$^{.4876}$
Ring B Installed	Flow (cfm) = 60.67 x (Fan Pressure in Pa)$^{.4955}$
Ring C Installed	Flow (cfm) = 21.37 x (Fan Pressure in Pa)$^{.5132}$

You can find this table on Page 57 of the TEC Model 3 Blower Door manual.

Note that the iTEC-700 app and wifi link has a *capture* feature that allows you to record the data of your operating point. This provides both the flow and the pressure

The best information on setting up blower doors is on the websites for The Energy Conservatory and Retrotec. It is best to know what you are doing rather than memorize a series of button pushes and hose connections. The point of a blower door test is to create a known pressure difference between the inside and the outside of the house so that the amount of air leakage can be measured. To accomplish that you need to know the pressure difference - the delta P - between the inside and the outside, the pressure inside with reference to (WRT) outside. Is the pressure inside higher or lower than the pressure outside? Common practice with a blower door is to extract the air from the house with the fan blowing out, putting it under negative pressure with reference to outside. The story goes that in the early blower door days in the cold climate of Minnesota they started out testing by pressurizing the house, the fan blowing in. In the cold weather that was brutal for the homeowners as well as the people doing the testing. Someone had the brilliant idea of depressurizing the house - turning the fan around. Seems pretty obvious from this vantage point, but it was a great new thought at the time.

And why 50 Pascals? Why not 30 or 90? With a pressure difference of 50 Pascals, the house is being subjected to the equivalent of a 20 mph wind on all sides simultaneously. That doesn't happen in real life - at least not on all sides of the house at the same time! At 50 Pascals all of the leakage is being created 'artificially' by the fan. That would happen at a higher pressure as well, but too much pressure can damage things. Too little pressure just isn't enough. 50 Pascals is the Goldilocks point in blower door testing.

There are situations where pressurizing the house is necessary. When the ducts are being tested under pressure, the blower door has to pressurize the house as well in order to equalize the pressure in the ducts with the pressure in the house to measure just the leakage to the outside of the pressure boundary. (If the pressure in the house is equal to the pressure in the ducts, there will be no driving force across holes that are inside of the house.) If there are pollutants in the house like asbestos or mold spores that would become airborne by depressurizing the house, it would be better to pressurize it. And there are occasions when a program or agency will require pressurization rather than depressurization testing.

It is also a good thing to understand how the manometer calculates airflow. Manometers measure pressure, but they can be programmed to calculate a lot of other things like airflow. Generally, knowing the pressure across a hole and the area of the hole is enough information to calculate the airflow through the hole in cfm. The hole sizes for the rings on the blower door are built into the software on the manometers. So Channel A displays the pressure and Channel B displays the airflow and has the capability of calculating other things like flow velocity.

Checklist: When a pilot is about to fly a plane, he/she uses a checklist to check out all the little parts (like the wings and the tail and wheels and the pitot tubes) to make sure they are all there, not damaged, and working properly. At least I hope the pilot does that! I hope he or she does it every time no matter how many thousands of hours they have flown that plane. You should have a checklist before you do a blower door test as well. And you want this to be a checklist that you actually use as a checklist not just some extra piece of paper. If you check off the items on the list with an actual check, you can put the list in the file with the other materials and know that you've done it. Here are some thoughts:

Check	Activity	Rationale
	Fireplace/woodstove covered	Sucking ashes out is messy and requires major clean up. You might not want to completely cover the fireplace opening. This is a major hole in the house. Covering the lower ¾ is probably enough.
	No loose soot in flues	There's always a white rug!
	Water heater(s) turned to pilot	Don't want the water heater to fire and have CO drawn down the flue during the testing.
	HVAC Thermostat(s) set to "off" & Fan to "Auto". Or switch off the Emergency service switch	Don't want the furnace to fire during the testing either. NOTE that you definitely want to remember to turn these back on before you leave.
	Exhaust fans turned off	They will be competing with the blower door for depressurization airflow and create an inaccurate reading.
	No loose drywall/plaster	You don't want to have to repair the building before you leave.
	Suspended ceiling tile displaced	Suspended ceilings are often installed to cover holes and other defects. Pulling the suspended ceiling down during a test can ruin your whole day.
	All interior doors open including closets	If you want to measure the leakage from the house as a whole, the air needs to be drawn out of all the rooms and spaces. Closets can have serious air leaks.
	All exterior windows closed (and locked)	You want to measure the airflow through the permanent holes not the adjustable ones.

	Window air conditioners in normal winter conditions	If the homeowner leaves in the air conditioner in place during the winter, it is a big hole. If they cover it or remove it, then they should cover or remove it prior to your test.
	Whole house comfort ventilators in winter condition	These are commonly called "whole house fans". Some of them are wide open all the time. Some of them have automatic covers. Make sure they are in their normal winter condition.
	Clothes dryer(s) turned off	Another competitor for the air in the house.
	Attic fans turned off	Unless the top floor ceiling is perfectly sealed, attic fans reduce the attic pressure below the house pressure and draw air from the house into the attic. (It's virtually impossible to turn off solar powered attic fans!)
	No vermiculite insulation	You don't want those particles floating around the house.
	No loose asbestos-like materials	You don't want this stuff floating around either.
	Pets secured	You will be leaving the door open. It ruins your whole day chasing Fido around the neighborhood.
	Bathtub surrounds	Plumbers put major holes under bathtubs. If the decorative surrounds are not properly secured, blower door testing can pull them off and they can be breakable.
	Stained glass secure	Stained glass windows can be delicate. Think about replacing that puppy!

There are three modes to perform the blower door tests: PR/PR (Pressure/Pressure), PR/Flow, and PR/Flow @50. Some programs require readings be made in all three modes.

Pressure/Pressure: As described above, if the pressure difference and the size of the hole is known, the rate of airflow can be calculated. By noting the ring being used on the fan and knowing the pressure difference across the fan a calculation of the airflow can be made. (See the table above.)

Pressure/Flow: This mode setting allows the manometer to automatically calculate the flow: displaying the pressure on Channel A and the flow on Channel B. This is how the house is actually performing at the time of measurement. When the fan speed increases, both the pressure and the airflow change.

Pressure/Flow@50: This mode is an automatic, convenience setting. Not all houses are able to be depressurized to 50 Pascals. The pressure that the house reaches, however, is an indication of the leakage if the house COULD reach 50 Pascals. The required 50 Pascal reading can be estimated from the maximum pressure that can be achieved. The accuracy of this reading depends on how close the house actually gets to 50 Pascals. It is tempting to use this mode for all readings, but the actual performance numbers for the house at any moment may not be what you think they are. For example, since you don't actually have to get to 50 Pascals, if you are doing any zonal pressure diagnostics, the house may not actually be at 50 Pascals and all the zonal readings will have to be adjusted to compensate for the actual pressure.

The Energy Conservatory (TEC) TEC has a free simulator for setting up the door.
http://products.energyconservatory.com/blower-door-simulator/

BPI QCI Field Test

- Combustion appliances set to pilot or disabled
- Blower door set up properly
- Manometer set up properly
- House set up properly
- Baseline measure and applied correctly
- Accurate CFM_{50} measurement
- Convert CFM_{50} to ACH_{50} (This is extra)
- Pressure pan testing
- Proper manometer setup
- Accurate measurement taken (Recognition of CFM@50 condition inaccuracy)
- Proper interpretation of the results
- Pressure diagnostics
- Measure zonal pressure differential to one appropriate zone - like the attic
- Proper interpretation of the results
- Measure zonal pressure differential to the basement/crawlspace (This is extra)
- Proper interpretation of the results
- Duct Pressurization Test
- Proper manometer set-up
- Pressure taps set appropriately
- Accurate measurement
- Fan Flow
- Accurately compared existing exhaust flow ventilation with rated capacity

CO Testing

Carbon monoxide (CO) is bad stuff. It kills people and it's not something to take casually. A personal CO monitor is a great investment. The sensors are like batteries unfortunately: they die after about two to three years. But if you are spending a lot of time in various houses, it's a good idea to continuously monitor the ambient CO. Monitoring it doesn't mean clipping the detector to your belt and forgetting about it, waiting for the automatic alerts to occur. It means looking at it periodically to evaluate the atmosphere you are working in.

Test ambient CO outdoors prior to testing indoors. The reason for this is that if there is an elevated level of CO outdoors, you can be pretty certain that any CO you measure indoors was generated outdoors and not by some combustion event indoors. Some personal CO detectors have the capability to be zeroed out. Personally I like to know what the level of CO is all the time. I don't want to zero it out. It's still in the air around me.

CO Measurements in the Flue (See Combustion Safety Guideline for more information):

There are CO measurements that you will need to make in the flues of the combustion appliances. Standards vary on what type of CO measurement should be made: Air-Free CO or As-Measured CO.

Air-Free Carbon Monoxide - A measurement of CO in an air sample or flue gas that takes into account the amount of excess air (oxygen,O_2) in the sample, incorporating an adjustment to the as-measured CO ppm value, thus simulating air-free (oxygen-free) conditions in the sample. Usually measured in units of parts per million (ppm).

As-Measured Carbon Monoxide – A measurement of CO in a sample of air or flue gas that does not take into account the amount of excess air (oxygen, O_2) diluting the CO concentration. Usually measured in units of parts per million (ppm).

(For a great list of standard definitions visit: http://www.karg.com/papers.htm)

During the combustion process, the level of CO in the exhaust gases change. A single measurement 'snap-shot' may provide false information. The level of CO should decrease somewhat as the system burns toward a steady-state condition. David Richardson in his guest post on the Energy Vanguard blog said it well (http://bit.ly/1FpOm33):

> *The CO numbers should never go up or continue to climb once the initial components have warmed up. If the CO numbers continue to climb, the equipment is operating in a deteriorating condition. This pattern of numbers tells us that the combustion process is breaking down inside the equipment itself. When rising CO numbers are present, combustion air is not making its way into the burner compartment and flue gases are not leaving the heat exchanger.*
>
> *"Rising CO numbers are one of the most dangerous and overlooked conditions in CO testing and cannot be discovered with a single-point CO test. The reason this condition is so dangerous is that you have no idea of where the CO numbers will stop. There have been many instances where rising CO numbers have gone from numbers that would look acceptable in a single-point CO test to very dangerous in mere seconds due to the deteriorating condition inside the equipment."*

Note that this testing process is different between the original (BA Standard) and the most recent BPI Building Analyst Standards (1100 & 1200). For the QCI testing you need to refer to the original guidelines, which prescribes a single point test. The BPI 1200 Standard that was recently adopted prescribes a multi-point test taken five times in one-minute intervals.

CO Action Levels

CO PPM	Time	Symptoms
9	24 Hours	National ambient air quality limit for CO. CO "Action" level for many public safety organizations.
30	8 Hours	UL 2034 CO Detector must NOT activate before 8 hours
35	8 Hours	Maximum 8 hour workplace exposure (OSHA)
70	1 Hour	UL 2034 CO Detector must NOT activate before 1 hour
150	10 min.	UL 2034 CO Detector must NOT activate before 10 min.
200	3 Hours	Mild headache, fatigue, nausea and dizziness
400	4 min.	UL 2034 CO Detector must NOT activate before 4 min.
400	2 hours	Serious headache, dizziness, nausea, Life threatening after 3 hours
800	45 min.	Dizziness, nausea, convulsions, and unconsciousness. Death within 2 – 3 hours
3200	5– 10 min.	Dizziness, nausea, convulsions, and unconsciousness. Death within 25 – 30 minutes
6400	1 – 2 min.	Death within 10 – 15 minutes
12,800	30 sec.	Death within 1 – 3 minutes

Note that UL 2034 Approved CO detectors must meet the UL requirements above but also:

- Must activate before 240 minutes at 70 PPM – the space could be at that level for 4 hours before the alarm goes off.
- Must activate before 50 minutes at 150 PPM – the space could be at that level for 2 ½ hours before the alarm goes off.
- Must activate before 15 minutes at 400 PPM

If the house is relying on a UL approved CO detector and it goes off, leave the house quickly!

Also point out to the homeowner that CO detectors have a life expectancy of 2 to 5 years. They should be tested regularly and it might advisable to buy the plug in type since they can be replaced more easily.

BPI QCI Field Test

- Tested ambient CO outdoors
- Tested ambient CO indoors
- Properly interpreted measurements

- Monitored ambient CO levels in the CAZ during entire combustion safety test

III.2.1 Heating System Diagnostics

Visual Inspection of flue system for problems

Check the flue for at least a quarter inch rise per foot. That's about 2 degrees. If you want to know, use a level if it's not obvious by eye.

Look for signs of condensation at the joints. The traces of white, chalky substance indicative of the minerals in the flue gases condensing. This may be an indication of inefficient combustion.

Note the proper location of the flues combined in a chimney. Some jurisdictions allow the exhausts from both oil and gas appliances to be inserted into the same flue. The lower BTU appliance should be entering the chimney above the higher BTU appliance.

If there is a thermally responsive vent damper in the flue, it should be replaced with a motorized vent damper or solid pipe. These devices are designed to make the appliance spill - at least for a short period of time.

Ascertain if the size of the flue is appropriate.

Identify Existing heating/cooling system component safety concerns:

Look for combustible materials in the vicinity of the combustion appliance - gasoline, propane, paints, rags, stacks of newspapers, etc. The homeowner – not by the QCI, should remove these.

Safety controls and switches should be in good operating condition such as thermal cutoffs, pressure switches.

Temperature/Pressure relief vents should be pointing down and approximately 6" above the floor.

Steady State Efficiency Test: Combustion analyzers have been designed for professional combustion technicians and consequently provide a great deal of information including steady state combustion efficiency. Pointing that number out to the proctor may be all that you need to do for the field test - depending on the proctor. But it is a good idea to know what it means.

The Annual Fuel Utilization Efficiency (AFUE) is a measure of a gas furnace's (http://bit.ly/1GdNSTZ) efficiency in converting fuel to energy, by projecting the average thermal efficiency for a complete heating season. A higher AFUE rating means greater energy efficiency.

The Steady State Efficiency (SSE) is calculated by allowing a furnace to warm up, (commonly for 20 minutes), then measuring the input BTUs per hour (BTUH) from the fuel and output BTUH released to the house in the form of heat . You then divide the output BTUH by the input BTUH. For example, a furnace that outputs 62,250 BTUs per hour once it has warmed up and requires 75,000 BTUH of fuel has an SSE of 62,250/75,000 = 83%.

Calculating Steady State Efficiency

"Investigation of High Efficiency Furnace SSE Measurements vs. AFUE"
http://homeenergyplus.wi.gov/docview.asp?docid=13851

1. Start the combustion analyzer after operating the furnace for five minutes with the probe inserted through a hole into the combustion intake pipe near the furnace cabinet. Record the intake air temperature at this time.

2. After the instrument start-up sequence is completed, switch the probe to a hole in the exhaust pipe near the furnace cabinet. If the analyzer has a separate thermocouple for intake air temperature, insert it in the hole in the intake air pipe.

3. Take SSE readings after 10 an additional minutes of furnace operation.

4. If the intake air temperature used by the analyzer is outside the range of 25 F to 55 F, adjust SSE according to the following formula: SSE adjusted= SSE indicated+ (40 - T_{intake})/30.

5. If adjusted SSE is less than AFUE – 2 percentage points, check for:

a. Exhaust temperature < 130 F

b. CO < 100 ppm

c. Temperature rise and gas pressures within manufacturer's specifications

d. Presence of condensate in drain line.

6. If any of the criteria in Step 5 are not met, fail the unit.

Heat Rise Delta T: As air moves through the air handler its temperature is changed. During the cooling season, it enters the air handler warm and is chilled. During the heating season it enters the air handler cool and it is warmed. If the air is moving too fast, the temperature won't be effectively changed. If it is moving too slowly, the system may overheat and the heating elements switch off, controlled by the high limit switch. The temperature rise of the air through the heating unit should be set by the furnace manufacturer. If the temperature rise is below the specified temperature, the heating unit blower is moving too much air and the speed should be reduced.

By using the temperature rise method the airflow across the device can be calculated. The BTU output of the furnace needs to be known and the temperature rise needs to be measured.

Drill a hole in the return air duct near the furnace and insert a temperature probe into the hole. Drill another hole in the supply air duct, out of the line of sight of the heat exchanger, and place a temperature probe in the hole. When the system has run long enough to reach steady state, the temperature difference between the two probes is the temperature rise or Delta T of the system. The airflow through the furnace as well as the air conditioning coil can then be calculated by using the equation:

$$CFM = \text{Btu Output}/(1.08 \times \text{Delta T})$$

Note that the same two holes can be used to measure the total external static pressure across the system. Insert the hoses from the Input and Reference ports on one channel of the manometer. Switch the pressure reading units to "inches of water gauge (iwg)". This should be done prior to air sealing the ducting as the system may only be working because of the duct leaks!

(Excellent reference on all things related to HVACR: Fundamentals of HVAC/R (http://amzn.to/1LUYkP5) by Carter Stafield and David Skaves published by Prentice Hall, 2010)

BPI QCI Field Test

- Visual inspection of flue system for problems;
- Identify existing heating/cooling system component safety concerns;
- Measure heating system flue gas CO during combustion safety testing;

- Conducted Steady State Efficiency test on heating plant;
- Accurately measure heat rise Delta T.

III.2.2 DHW Diagnostics

Domestic Hot Water or DHW is certainly a component of comfort. Many DHW systems are set at 140 °F, which will scald a child within 1.5 seconds or an adult in 5 seconds. To measure the temperature of the hot water, you will need to allow the water to run from a tap near the water heater until the temperature stabilizes. The length of time will depend on the system in the house but you should allow five minutes. Run the water into a small container so that your thermometer can be completely immersed when you make your measurement. If the temperature is above 120 °F, recommend that the setting be lowered.

DHW systems come in a variety of flavors from the standard forty or sixty gallon tanks with Category I atmospherically vented combustion elements to instantaneous gas and electric systems some of which are activated when motion is sensed. Each one of these systems requires different checks. Verifying the temperature of the water they produce is obviously common to all since that is their primary purpose.

For the Category I, holding tank systems, the combustion safety checks need to be made - checking for spillage at the draft diverter, measuring the CO in the flue, measuring the draft in the flue. (Remember the maximum draft pressure in the flue with reference to the CAZ is calculated by dividing the outside temperature by 40 and subtracting 2.75 from the result. (T_{out}/40)- 2.75) These tests should be made under worst case conditions when all the devices in the house are drawing out as much air as possible, making it as hard as possible for the water heater to successful push the combustion gases up the flue. Before performing the combustion tests, the safety equipment on the appliance should be verified, particularly the Temperature/Pressure relief valve, which should not be blocked, should be pointing down toward the floor, and terminating about six inches above it.

If you put your hand on the outside of the water heater tank and it feels warm but not hot, the tank should not need to be insulated further. Tank insulation can sometimes cause more harm than good particularly on gas water heaters. The insulation may slide down and block the intake air into the combustion chamber. It may block access to the thermostat(s). Most manufacturers do not recommend wrapping insulation around fossil fuel combustion appliances.

Tankless coils are commonly inserted into an oil burner. They are efficient during the heating season and they don't require an additional tank that consumes floor space. However, during the warm weather, the boiler has to fire just for hot water. They are controlled by an aquastat that prioritizes domestic hot water over space heating. During the cold weather, the Aquastat may be set to keep the water temperature as high as 180 to 200 degrees. Setting it back to 120 °F during the mild weather will save a significant amount of fuel. The most a QCI should do is ask the homeowner if the Aquastat gets reset in mild weather and/or suggest that they have a qualified heating technician do it or show them how to do it.

DON'T MESS AROUND WITH THE AQUASTAT UNLESS YOU KNOW WHAT YOU ARE DOING. CHECK WITH A QUALIFIED HEATING TECH.

Tankless water heaters: Tankless hot water heaters need to be flushed about every 12 to 18 months to prevent scaling. This is not something that a QCI needs to provide, but it would be helpful for the homeowner to know that it should be a part of their regular maintenance chores. In checking out the system, check to make sure that the plumber installed flush-out hose fittings so that regular maintenance can be performed. (A useful video on the subject from Matt Resigner: http://bit.ly/1D6Xc4e)

Installation checks for a Category III tankless water heater:

- ✓ **Improper Venting:** The most common error is the use of standard B-vent, which can quickly destroy the heat exchanger due to acidic condensation. Improper venting may cause exhaust gas leaks and would be considered unsafe. All non-condensing tankless water heaters must use sealed stainless steel vent pipe approved for use in Category III appliances. Most condensing models may use Schedule 40 PVC, ABS pipes, or Category III stainless steel.

- ✓ **Gas Supply Line Problems:** Undersized Line: There is a great tendency to just "hook up to the existing gas line," which is usually only ½ inch diameter. Strict attention must be paid to the gas-line requirements section in the installation manual. Make sure that you size for the maximum BTU per hour rating and not the minimum.

- ✓ **Make sure that it is the correct heater for the type of gas, Natural Gas or Liquid Propane.**

- ✓ **Water Flow/Temperature Fluctuation Problems:** If there is low flow from the heater, the first thing to check would be the input filter. Turn off the cold supply, relieve the hot pressure and remove the filter to check for debris. Clean or replace the filter and then check the flow again. If there is inadequate hot water or variations in temperature, check for plumbing "cross flow," which means there is a connection somewhere between hot and cold. This may be as inconspicuous as a worn seal in a single-handle faucet. To check for this, shut off the cold supply to the heater and open either the pressure relief valve or any hot-water faucet. If there is a steady flow of water from the hot output after the pipe has drained, there is a cross flow somewhere in the plumbing that must be fixed.

✔ Other causes of temperature fluctuations can be the gas line may be too small, or the heater temperature may be set too high in an effort to "get more hot water." This can cause a shower to be turned too far to the cold side, which doesn't allow enough hot-water flow to keep the heater on steadily. In most circumstances, the heater should be set to its default of 122°F.

✔ **Hard Water Problems:** In areas where the water is very hard with high levels of calcium and magnesium (about 7+ grains per gallon on residential application), manufacturers recommend that the water be treated with a water softener or some other descaling device. Do not use a reverse osmosis process! It will destroy the heat exchanger. Removing minerals or keeping them in suspension will greatly lengthen the life of the heater and reduce service problems.

✔ **Environmental Issues:** The heaters should be installed in relatively clean areas with access to sufficient combustion air that is free from dust, lint, grease, or chemicals. If any of these conditions are present, a direct vent model or kit should be installed. Remember that these units are drawing their combustion air from their surrounding environment. For this reason, restaurant, beauty salon and industrial applications should always get a direct vent unit, which draws its combustion air from the outside.

Information from Takagi http://bit.ly/1PRqsWB

BPI QCI Field Test

- Measure the temperature of the domestic hot water at the tap;
- Check for worst case spillage in DHW;
- Perform worst case draft test on DHW;
- Measure DHW flue gas CO;
- Apply BPI action levels based on test results for CO in the flue.

III.2.3 Oven Testing

I have been compelled to leave more homes because of the CO generated by the oven than any other cause. It is common for ovens and stoves to be used as a heating element. If they are expelling CO into the home, they can be deadly. They have no chimneys and a high percentage of them only have recirculating range hoods that don't vent to the outside. These hoods are designed to remove the odors from the effluents and that only happens when their filters are new and clean. Recirculating range hoods are like recirculating the water in your shower!

Anything that burns can create CO. Although the primary testing protocols are for gas ovens, even electric ovens can create CO - even microwave ovens and toasters can if something burns in them. Before you begin testing the oven you need to be sure that there is nothing in it. People have a tendency to store pots, pans, crackers, bread, and other combustibles in the oven. (The thought is that the warm oven is like a brisker and will keep food from going stale in a damp climate.) Setting the house on fire is definitely not a positive Quality Control procedure! So look in the oven before you crank the temperature up to 500 °F. Once you do turn the temperature up, hang around. It's going to take some time to warm up, but it is better to be there than somewhere else in the house if anything goes wrong. Use the time to test the burners, checking out their flame patterns and CO generation. Locate the opening to the small vent from the oven so that you will be able to insert the probe from your combustion analyzer.

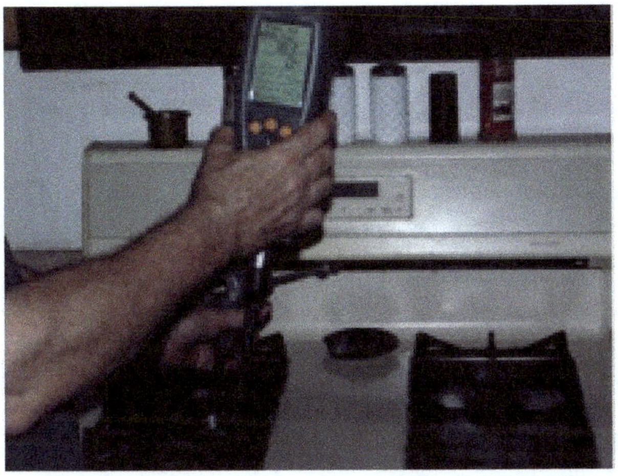

Monitor the ambient CO during the entire test. Since an elevated level of CO is going to be delivered into the kitchen during this test, it is a good time to set a CO alarm from the house on the stove surface. CO alarms have about 3 year lives. Remember that the sensor is a bit like a battery - once it's life is over it should be disposed of and replaced. I sometimes put the household CO detector on the stove top along with my personal CO monitor (which is hopefully regularly checked) and see if the household detector responds . . . at all.

Once the oven reaches steady state - generally after about 5 minutes - insert the probe into the vent and measure the CO in the exhaust gases. At that point the CO level should remain steady. If it is still going down and it is below 100 ppm, the oven has passed the test. If it is going up, then the oven should be adjusted. It is difficult to know who you can call to tune an oven. You can't look up oven "Oven tuning Services" in the yellow pages! I have brought in appliance service technicians who don't have CO detectors or monitors and really don't know what I am talking about when I tell them that the oven is producing an elevated level of CO. So if the oven in the house you are testing does have an elevated level, the BPI Standard says to provide exhaust venting, but the homeowner should be advised of the issue. It is one of those teaching moments where the local appliance supplier could be provided with a procedure for tuning an oven, reducing the CO level, helping his/her customers, and providing a public service.

Oven tuning information from A. Tamasin Sterner, Pure Energy Coach,
http://www.pureenergycoach.com/ via Pam Denlinger at Solaire Energy
http://www.solairenergy.com/

Clean an oven burner:

- Turn all the appliance controls to *off*
- Remove the oven bottom (typically held in place by tabs or fasteners).
- Remove the baffle beneath the oven bottom.

- Turn on the oven and watch the burner. If the flame is not continuous along the length of the burner, some flame ports may be clogged. Turn off the oven and let the burner cool, then clear each port in the burner with a sewing needle.

Fix-It Tip

What should a well-adjusted gas oven (http://bit.ly/1SOmsHU)flame look like? It should be blue in color, symmetrically shaped, and about 1/2 inch tall. A ragged, hissing flame indicates too much air. A yellow orange flame indicates too little air. Before adjusting, make sure the oven is turned off and cool.

Adjust an oven flame:

- Turn all of the appliance controls to *off* Open the door and remove the oven bottom and baffle (see above).
- Turn on the oven and inspect the flame.
- Identify the air shutter that will be adjusted. It may be at the base of the oven burner, just above the safety valve, or somewhere along the gas delivery tube.
- Loosen the setscrew that locks the shutter or plate, and adjust the opening to increase or decrease the amount of air mixing with the gas.
- Turn the oven on again to verify the adjustment, repeating as needed.
- Once the flame is correctly adjusted, turn off the oven and tighten the setscrew.

Cleaning the Burners

Clogged burners are a very common problem with gas ranges because foods spilled on the burners block the gas ports and prevent ignition. On some gas ranges you can remove the top ring of the burner to expose the ports. Here's how to clean a burner:

Step 1: Turn off the power supply, both gas and electric, to the range. Then remove the burner.
Step 2: Soak the burner in a solution of mild household detergent and water. Clean it with a soft cloth.
Step 3: Clear the gas ports with a pin or needle, rinse the burner, and let it dry. Caution: *Do not use a toothpick or matchstick to clean the gas ports. If the tip of the wood gets stuck in the burner ports, it could cause a serious blockage.*
Step 4: When the burner is completely dry, replace it, and turn on the power and the gas supply.

Gas Oven: Pilot Lights, Shutoff Valves, and Ignition Systems

The flame of a gas range burner should be steady and slightly rounded, with a light-blue tip. To adjust the flame, slide the air shutter mixer plate open or closed.

The flame of gas range burners should be steady and slightly rounded, with a light-blue tip. The flame should be quiet and should respond to adjustments made at the control knobs. Most burner troubles can be quickly solved by adjusting the air shutter mixer plate, which is located at the end of the burner tube near the knob controls. Turn a small screw on the plate, slide the plate open or closed, and tighten the setscrew.

If the flame is yellow, it's not receiving enough air. To allow more air in, open the plate slightly. If the flame is high or makes a roaring noise, it's getting too much air and you should close the plate slightly.

Servicing Pilot Lights

The oven pilot is usually located at the back of the oven; turn a screw on the pilot ignition unit to adjust the pilot flame height.

One pilot light usually serves all the top burners of a gas range. Some ranges have two pilot lights, one for each side of the range. A correctly adjusted pilot flame is steady and blue, between 1/4 and 1/2 inch high. If the flame goes out repeatedly, or if it's yellow at the tip, it's getting too little air. If there's a space between the flame and the pilot feed tube, it's getting too much air. To correct either condition, turn the pilot adjustment screw on the gas line slightly, as directed by the manufacturer of the range.

Most modern gas ranges use an electronic pilot igniter to start the pilot.

If the pilot flame is properly adjusted, but the flame doesn't ignite the burners, the problem is probably in the flash tubes that run from the pilot to the burners. These tubes may be blocked by spilled food from the burners. If this is the case, turn off the power to the range and clean out the tubes using a short piece of wire. Push the wire through the opening until the tube is clear. You may have to disconnect the tube to clear it. After cleaning the tube, replace it in the same position.

If the pilot has a switch, the switch may be faulty. Turn off the power to the range and test the switch with a volt-ohm-milliammeter (VOM) set to the RX1 scale. Remove the retaining screws that hold the switch in the range cabinet, and disconnect both electrical lead wires to the switch. Clip one probe of the VOM to each switch terminal. If the switch is functioning, the meter will register zero. If the meter reads higher than zero, the switch is faulty and should be replaced. Remove the switch and replace it with a new one of the same type. Connect the new switch the same way the old one was connected.

Most newer gas ranges and ovens don't have pilot lights. Instead, the gas is ignited by an electric ignition system. In this type of system, an element becomes hot and glows like the filament in a light bulb when an electric current passes through it. The heat from the filament lights the gas. As a rule, these ignition systems are sealed and cannot be repaired or adjusted. When an electric ignition device fails, don't try to fix it. Call a professional service person for replacement.

If the flame is set too low, adjust it. Next to the pilot, locate a small boxlike unit with a couple of screws in it. This is the ignition unit. The ignition unit could also be located below the pilot; follow the gas line down until you locate it. Turn one of the screws on the ignition. Experiment, turning the screws a little at a time, until the flame is adjusted properly. It should not be as high as the top pilot flame; leave it as low as possible. On some oven pilots, turn the control to the OFF position and light the pilot; then turn the oven dial to the BROIL position. The pilot will heat the controls to the ignition switch.

Servicing Automatic Shutoff Valves

On some ranges, there is an automatic shutoff valve located in the pilot assembly that shuts off the gas to the burner any time the pilot and the burner are both off. If this unit malfunctions, don't try to fix it yourself; you should call a professional service person for repair or replacement.

An electrically operated shutoff valve used on some gas ranges has two facing valves, an electromagnet, and a manually activated reset button. The thermocouple fitting is next to the pilot valve, as in most burner systems. A small amount of electricity holds the facing valves apart. If the pilot light goes out, no electricity is generated, and the valve closes to turn the gas off. On this type of system, relight the pilot by depressing the reset button and holding a match to the pilot. It should take about a minute to light this pilot. If you aren't able to relight this system, call a professional service person.

Another note about CO alarms: UL approved CO alarms are meant to offer protection to healthy adults during very high levels of CO in the home. A UL listed CO alarm will allow you to breathe:

30 ppm for up to 30 days;
70 ppm for up to 4 hours;
150 ppm for up to 50 minutes;
400 ppm for up to 15 minutes

BPI Building Analyst Standard (v1/4/12):

For gas ovens, CO shall be measured at steady state (usually after 5-10 minutes of operation) at the highest setting. When measuring CO on gas ovens, it is recommended to turn on the exhaust hood and open a window to reduce risk of exposure to elevated ambient CO levels.

Ambient CO levels shall be monitored upon entering the combustion appliance zone and during the test period for all appliances. If ambient levels exceed 35 ppm at any time, turn off the appliance immediately and make appropriate repair recommendations according to the charts provided.

Range tops and ovens produce moisture and oxides of nitrogen. Excess moisture is not good for the durability of the home (possibly contributing to mold problems) and NOX is not healthy. These combustion appliances are capable of producing CO, which is a health hazard. In all cases a carbon monoxide detector is recommended and homeowners should use exhaust ventilation when using these appliances. New appliances may require an extended warm up period to reach steady state.

1. *Remove any items/foil in or on oven.*
2. *Make sure self cleaning features are not activated, set oven to highest setting.*
3. *Test oven for CO in the flue, before dilution air.*
4. *After 5 minutes of operation, check for steady state'*

Level I Action - 100 ppm to 300 ppm as measured you must install a carbon monoxide detector and recommendation for service must be made to the consumer.

Level II Action - Greater than 300 ppm as measured - the unit must be serviced prior to work. If greater than 300 ppm after servicing, exhaust ventilation must be provided with a capacity of 25 CFM continuous or 100 CFM intermittent.

**Continually monitor ambient CO levels during test.*

Since all gas cooktops generate CO and it is difficult to simulate an actual operating condition for these appliances during the course of a typical house inspection, specific action levels for these burners are not specified by BPI. However, technicians must specify appropriate measures to mitigate potentially dangerous CO production of these units. ASHRAE exposure limits for CO shall be referenced when making recommendations for CO control in these areas. The recommended ASHRAE limit for 24-hour exposure of 9 ppm shall be applied to building occupants. In most cases, it will not be possible to effectively test for this condition, however the following measures shall be recommended whenever gas cooktops exist in the home:

If burners do not ignite properly or do not burn cleanly, a clean and tune of the appliance shall be recommended.

If the appliance is located in a confined space and mechanical ventilation is not readily available, mechanical ventilation shall be recommended.

Ventilation provided for unvented gas ovens must provide a minimum capacity of 25 cfm continuous airflow or 100 cfm intermittent.

III.2.4 Duct Diagnostic Testing

An HVAC duct system is dedicated to conditioning the air IN the house. Conditioned air is pushed out of the air handler, through the ducts, through the rooms, drawn back into the return registers, and back to the air handler where it is reconditioned (heated or cooled) to repeat the cycle. Unless there is an intentional connection to the outside for fresh air, there should not be any connection to the outside! It is not a ventilation system. It shouldn't add or subtract air to or from the house. In operation, it should not change the pressure in the house. If the system has been designed and installed properly, the amount of air leaving the air handler should equal the amount of air returning to the air handler.

Since the air is passing through the rooms as it is passing through the ducts, the rooms themselves are part of the duct system. And if a room only has a supply, then the door to the room becomes a damper. Closing the door will cause that room to increase in pressure relative to the house. At the same time, less air will be able to get back to the air handler and so the rooms where there are return vents, will have less pressure which might cause unintended consequences such as drawing humid outside air through holes in the walls or ceilings where the moisture might condense, degrading the building.

For the system to work properly, the ducts need to be sealed tightly. If there are leaks on the supply side outside the pressure boundary, less air is delivered to the house than is subtracted from it, putting the house under negative pressure. The opposite is true if there are leaks on the return side.

NOTE: If duct leakage testing is not thoroughly familiar to you, you need to practice the process prior to taking the QCI field test. The equipment product manuals from The Energy Conservatory and Retrotec are excellent. You should spend some time with them. Testing the ducting is like testing a room in a house. If all the intentional openings in the ductwork (i.e. the grilles and registers) are sealed, the only leakage will be through the remaining unintentional openings. So to perform the duct test, the intentional openings need to be sealed with tape or grille mask or some other tight covering, the duct tester fan needs to be connected to a large return opening or the air handler itself, and the ducting pressurized or depressurized. The airflow through the fan will be the airflow through the unintentional leaks and cracks in the system.

Why are ducts tested at 25 Pascals of pressure? HVAC design of ducting is commonly set up to pressurize the ducts to 0.1 iwg (inches of water gauge). 25 Pascals is a similar pressure and the ducting should be capable of performing well under that pressure because that's the pressure that it will experience ever day that it runs!

Total Leakage Testing:

Ducts carry air to and from the air handler. It is common for some of the ducts to be outside the air and/or thermal barrier. A Total Leakage test is a measure of all the leaks in all the ducting no matter where they are. If there are multiple systems in the house, they all have to be tested and their leakage combined.

Some states and programs have a variety of duct testing requirements.

To comply with the IECC 2012, only systems with ducting outside the thermal envelope need to be tested. If it is not clear that all the ducts are inside the thermal envelope, duct testing should be done.

2012 IECC: *Duct tightness shall be verified by either of the following:*

1. Post-construction test: Total leakage shall be less than or equal to 4 cfm (113.3 L/min) per 100 square feet (9.29 m^2) of conditioned floor area when tested at a pressure differential of 0.1 inches w.g. (25 Pa) across the entire system, including the manufacturer's air handler enclosure. All register boots shall be taped or otherwise sealed during the test.

2. Rough-in test: Total leakage shall be less than or equal to 4 cfm (113.3 L/min) per 100 square feet (9.29 m^2) of conditioned floor area when tested at a pressure differential of 0.1 inches w.g. (25 Pa) across the system, including the manufacturer's air handler enclosure. All registers shall be taped or otherwise sealed during the test. If the air handler is not installed at the time of the test, total leakage shall be less than or equal to 3 cfm (85 L/min) per 100 square feet (9.29 m^2) of conditioned floor area.

***Exception**: The total leakage test is not required for ducts and air handlers located entirely within the building thermal envelope.*

Leakage to Outside

If there is duct leakage inside the thermal boundary, the anticipated amount of conditioned air might not be delivered to where it is supposed to be and the house might not be as comfortable as it should be. However, from an energy efficiency point of view, it is the losses to the outside of the house that are a waste of energy. The system is attempting to condition the great outdoors, and that would be tough to do!

To eliminate the leaks inside the house, during testing the blower door is used to put the house at the same pressure as the pressure in the ducts. When there is no difference in pressure to move air through a hole, there is no air movement. The test needs to be set up so that the ducts outside the thermal envelope are clearly outside the thermal envelope - opening the bulkhead door or crawl space window, for example. Remember that if the ducts are being pressurized, the house will also need to be pressurized. I like depressurizing the ducts for to test the leakage because it draws the tape or grille covering down more tightly on the grilles rather than blowing it off. That way I can leave the blower door in its normal configuration - depressurizing the house.

QUESTION: How could duct design and installation cause a forced air system to shut down during the heating season?

A. Too much leakage not allowing enough air to flow through the system.

B. Not enough air flowing through the system, overheating it and causing it to shut down on the high limit control;

C. Too much air flowing through the system. Reduce the blower speed;

D. Thermostat is set too high.

(ANSWER: B – Occasionally the only reason a heating system works consistently is because of the leaks in the ducting. By sealing the ducts without measuring the pressures in the system, the amount of air flowing by the high limit control can be restricted to the point where the system shuts itself down.)

——————————————

BPI QCI Field Test

- Duct Pressurization Test
- Properly set up the manometer
- Pressure tap appropriate
- Accurate measurement
- Pressure Pan Test
- Properly set up the manometer
- Accurate measurement taken
- Properly interpreted the results of the pressure pan testing

III.2.5 Mechanical Ventilation Testing

[There is a Ventilation Systems Testing Protocol Document at the end of this book. It also includes a description of how to test range hoods. This is one of the requirements in version 3 of the ENERGY STAR new homes program. The HVAC Quality Installation Rater Checklist requires the home energy rater to measure the airflow of the range hood to ensure that it's pulling at least 100 cubic feet per minute.]

The fact is that new houses are getting built more tightly: it's in the codes. It is also a fact that weatherizing a house also increases the air tightness. Mechanical ventilation is simply a fact of life. For the BPI QCI test it is one simple measurement, but for more general quality control you must know about bit more about it.

ASHRAE (American Society of Heating Refrigerating and Air Conditioning Engineers) has a Standard (62.2) that defines the national, residential ventilation requirements. It includes sizing for whole building ventilation as well as local exhaust. It also includes a lot of other information and is worth checking out. (ASHRAE Standards http://bit.ly/1eJW1n3) Since mechanical ventilation in homes is necessary and a code requirement, why not take the advice of a group of people who spend their lives thinking about residential ventilation? They always think about it. They study it. They argue about it, and they struggle with every word. The Standard certainly isn't perfect, but there is a lot of thought involved in it.

Building Airflow (Ventilation) Calculation

Since the original BA Standard is used for the HEP QCI test, the ventilation calculation is based on using the BAS or Building Airflow Standard calculation. The process is pretty well explained in the Standard, but it is a good idea to know why you are making the calculations.

Step 1: Calculate the Ventilation Required for the Building

This is a calculation of the volume of the house including the basement. When this Standard was written, an acceptable natural airflow through the building was about 0.35 air changes per hour – a change of about 1/3 of all the air in the house every hour. To figure that out, the volume of the house needs to be calculated and multiplied by 0.35. And since ventilation systems are marketed at cubic feet per minute (cfm) ratings, the resulting number is divided by 60 – the number of minutes in an hour.

Step 2: Calculate the Ventilation Required for the People

The reason for the ventilation system is to make sure that the people living in the house are healthy so the Standard adds a verification calculation for the people, providing 15 cfm per person. (If you don't know the actual number of people living in the house, you can use the number of bedrooms, adding an extra person for the master bedroom (number of people equals the number of bedrooms plus one).)

Step 3: Using the Higher Airflow Requirement, Convert to CFM_{50}

The higher result of the calculations in Step 1 and Step 2 can then be used to determine a limiting CFM_{50} number below which the house will be too tight to theoretically leak enough air naturally, without an additional mechanical ventilation system.[6] Calculate the limiting CFM_{50} value by multiplying the higher of Step 1 or Step 2 by the N factor for your location:

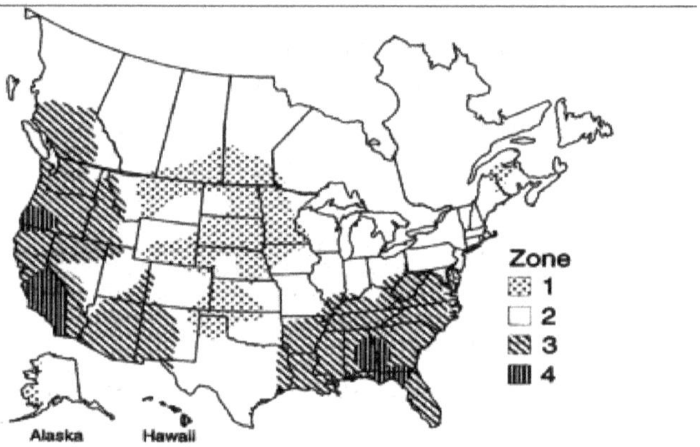

Zones	1	2	3	4
Stories				
1	16	19	22	25
1.5	14.2	16.9	19.6	22.3
2	13	15.4	17.8	20.3
2.5	12.2	14.4	16.7	19
3	11.5	13.7	15.8	18

Example:

Living space = 1,700 ft^2

Basement = 700 ft^2

Volume = 8 x (1,700 + 700) = 19,200

\# of Bedrooms = 3

People estimate = 3 + 1 = 4 people

---

[6] Note that since the Standard was written, it has become clear that the air quality in a house can't rely on natural leakage if it meets the BAS calculations. Except in the leakiest of houses mechanical ventilation is required.

# of Stories above grade = 2
Location = Nashville, TN
N_{factor} = 17.8

Air Changes for the Building
$Airflow_{BLG}$ = (0.35 x volume)/60 min
$Airflow_{BLG}$ = (0.35 x 19,200)/60 = 112 cfm

Ventilation for the people
$Airflow_{people}$ = 15 x 4 = 60 cfm

112 is greater than 60

Minimum CFM_{50} = 112 x 17.8 = 1993

Most of the time the cfm for the building will be higher than the cfm for the occupants, but doing both calculations is good practice. Note that if there is a question on the written/on-line exam about the BAS, the values in the table listed here may be different than the values calculated from the BA Standard. You need to select an answer that is reasonably close.

The BAS calculation is being phased out as codes, standards, and programs move toward the acceptance of the latest versions of the ASHRAE 62.2 Standard. The latest version allows the required ventilation system volume level to be adjusted by the leakage of the house as measured by the blower door test. These calculations do not have a tightness threshold: tighten the house as much as your program allows and then install a mechanical ventilation system.

Once you know how much air you need to move (refer to the Residential Energy Dynamics http://bit.ly/1bQxsTP website to do this easily), you need to decide how to do it. It could be as simple as an exhaust fan in a central bathroom or it could be as complex as an energy recovery ventilator (ERV). The ASHRAE Standard doesn't prescribe the system to be used. Local exhaust is also required in the bathrooms and the kitchen. There also needs to be some sort of control, which can be as simple as an ON/OFF switch, or as complex as a microprocessor control that senses the outside temperature and/or the quality of the air in the house.

To comply with the Standard, the fan or fans need to be tested and rated by the Home Ventilating Institute (HVI) who rates ventilation products for airflow, sound level, and power consumption. The airflow rating is done at a very low pressure to provide a standardized benchmark that all products can achieve. The rating on the box will not be the performance of the installed product. The length, size, type, and route of the ducting that is attached to the fan will have a major impact on the amount of air that it moves. Even the hood on the exterior of the house has a significant impact. So the only way to know how much air is truly moving through the fan is to measure it.

Some airflow measuring devices were brought over from commercial airflow measurement that is significantly greater than the 20 or 30 cubic feet per minute (cfm) that is moving through a household bathroom fan. New products have been developed that work well. (A Ventilation System Testing Protocol is included at the end of this book. Note that this is not an official document. It is just an attempt to get us all on the same page for testing results.)

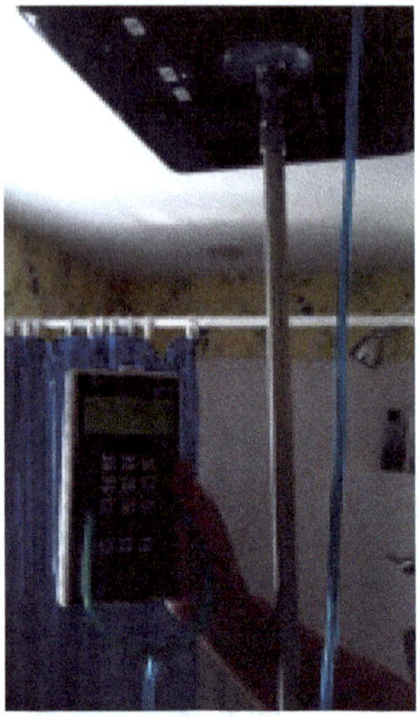

Box with an adjustable hole: The simplest of these is a box that has a hole of a known size and a digital manometer. By knowing the size of the hole and the pressure difference across it, the volume of airflow can be calculated. The Energy Conservatory (TEC) has a product that does this well. (http://bit.ly/1P1Vlao) There is a tool on the RED website mentioned above that will also do the calculation for you (http://bit.ly/1NDvW3k)

Large vane anemometer: Air moving across a vane will cause it to spin. The rate of spinning is proportional to the amount of air moving across it. Pin wheels are a good example! The rate of rotation, however, is a measurement of the velocity of the airflow and not the volume. In order to determine the volume, the area of the opening must be calculated. Testo makes a good one of these (http://bit.ly/1bQyK12) that also has an accessory capture hood that simplifies the process.

Hot wire anemometer: Air moving across a heated wire will cool the wire. The rate of cooling is proportional to the rate of airflow. Hot wire anemometers also measure the velocity of the airflow (fpm) and need to be converted to volume of airflow by calculating the area of the opening.

Balometer: A balometer consists of a capture hood (usually made of cloth) and a flow grid connected to a display. As the air moves across the grid inside the hood, the cfm is calculated and displayed. These devices are usually not suitable for residential applications.

Powered flow hood: Anything that is in the air stream will impact the performance of the fan and will affect the measurement. Several devices include a balancing fan. As the air moves through the system, the balancing fan pushes the same amount of air the fan being measured does. When the two flows are equal, the flow through the reference fan will equal the flow through the fan under test and since the reference fan is calibrated, the airflow through the system can be determined. A duct tester can be used for this or a modified duct tester such as the TEC Flowblaster (http://bit.ly/1BTh6zW) or a device called the Flowfinder2 (http://bit.ly/19LUsSq).

Note that just like the air moving from the fan to the outside impacts the performance of the fan, the air moving into the fan will also impact the performance: one cubic foot in equals one cubic foot out. If the bathroom door is closed, the flow through the fan will be different than if it is open. It the bathroom window is open, it will also have an impact. At this point there is no protocol about how the house should be set up to measure the flow through the fan.

The reading from any of these devices needs to be compared to what is on the work order or to the calculation of the required airflow from the ASHRAE 62.2 Standard or from the building codes. Measure it and compare it - that's all that is required for the BPI QCI test.

BPI QCI Field Test

- Measure mechanical ventilation performance
- Compare measured performance with rated capacity and work scope requirements
- Record results
- Accurately compared existing exhaust flow ventilation with rated capacity.

Task III.3 Identify Work Problems

Like actually finding a parking place in Manhattan - NYC - if you don't find work problems, you have to believe that you've missed something! There are a lot of items on the QCI field test related to "Work Problems". Since not all of them will actually happen during a field test, you will need to explain to the proctor what you would do if you did find them. Remember the word "document" . . . "if I saw this I would document it and then . . .." Remember to mention what you would DOCUMENT during the field test as well.

Comparing missing items to the work order is straightforward. If they were supposed to get done and they didn't get done, they should be done. This blends clearly into the issue of non-conformance – in this case non-conformance with the work order. Of course, there is also non-conformance with codes and standards and safety regulations and worker professionalism.

These steps can be performed throughout the field test rather than in a big slug all at once, but you will want to assist the proctor by identifying the specific items that he/she will be checking off.

BPI QCI Field Test

- Candidate reviewed the work in relation to the work plan;
  - Noted audit discrepancies (punch list)
  - Damage caused by workers
  - Documented non-conformance
- Candidate identified the need to run diagnostic tests;
- Candidate checked for missing opportunities;
- Candidate determined if the problem was material or work related;

- Candidate reviewed the results of the visual/sensory inspection;
- Candidate reviewed the results of diagnostic tests;
- Candidate compared diagnostic results with the post-test data;
- Candidate made a determination of pass or fail based on the results;
- Candidate noted the work problems if the result was a fail.

Domain IV: Ensuring Worker Professionalism

Task IV.1 Perform Spot Checks

If you have to check everything, you might as well just do the whole job yourself! And some people feel that way - that the only way to make sure that the job is done right is to do it yourself. There is an art to delegating. It was Ronald Reagan who loved the Russian proverb, "Trust but verify!" Since you can't check everything, how do you know what to check that might be the tip of the iceberg of a problem?

Start with the work order. Health and safety is the most critical element so you want to start there. You want to verify that common sense safety practices are being followed. It's not hard to tell if workers are on the roof without safety harnesses or blowing insulation without PPE. And then there are the work order items in the crawl spaces and the dark recesses of the attic. Just because you don't see duct mastic on one duct joint, does that mean that they didn't use it on all of them?

For example, you test the ducting in a house and it fails miserably despite the fact that the contractor swears that he sealed up every joint to the best of his knowledge and ability. The ducts are covered with insulation and the homeowner and builder are waiting for the certificate of occupancy. You put the fogger on the system and smoke is pouring out everywhere. The contractor comes back, grumbling, seals the ducting (when you're not there), and has someone else come in and test the system. It passes. Since no codes or standards say you have to retest the ducts, do you trust the contractor's test or do you test it again yourself?

A. You accept the results of the contractor's test and move on;
B. You accept the contractor's results because you have worked with him before and you trust him;
C. You retest the system because you've no personal knowledge of the contractor's craftsmanship.

The right answer to this question is C if you have any doubts about the results. Being sure that the ducts have been tested properly is the right thing to do. As the QC inspector it's important that you honor your own integrity. If you do retest the system and it fails again, what's your next move?

Some spot check decisions are based on knowledge of the crew and the quality of the work that they do. You're more likely to trust a crew that you have consistently worked with before and who have proved that their level of quality work is high. But at some point you still need to verify. How do you determine where that point is? Some issues have to be based on your experience and your common sense. There is always a clock running in the background: there is never enough time to do all the things that you might want to do. You don't have to pull up every piece of insulation in the attic to verify that all the gaps and holes were filled. (The fact is that if the work around the attic hatch is poor, the rest of the job is almost guaranteed to be poor as well.)

The same thing is true for reviewing the job-site documents. You can't read through them all. The energy auditor used his/her time, skill, and knowledge to test and evaluate the house and write up the audit and work-order. The crew chief used his/her time, skill and knowledge to get the right materials, assign the tasks, and oversee their implementation. All you should have to do is make sure that there are blower door results and they make sense, combustion safety results that meet the safety criteria, insulation bag counts and usage that verify that the amount used met the required levels, and that the paperwork has been filled out and written clearly.

It goes back to the need for checklists and filling them out every time. The point is to verify that the task has been done. But not just done . . . done well.

BPI QCI Field Test

- Mentioned need to perform spot checks on in-process work
- Mentioned performing random sampling of job site documents
- Mentioned random sampling of worker credentials
- Mentioned need to observe the workers while they are performing work
- Mentioned need to interview workers at the job site

Task IV.2 Provide Feedback Regarding Professionalism

Feedback is when the output of a system is conducted back as inputs for the initiation of the system to improve (or degrade) the performance. When you flip on a light switch and you see the light go on, that is instantaneous feedback. When a rat gets a piece of cheese for running through a maze, that's feedback. Feedback that enhances the desired or beneficial performance of the system is positive feedback. A reward for doing something right or well is positive feedback.

Feedback can also degrade the performance - negative feedback. If the rat were given an electric shock when it ran through the maze incorrectly, that would be negative feedback. That sort of feedback is designed to amend undesirable behavior.

If a homeowner has suffered with poor workmanship or bad results, they may be conditioned by negative feedback from believing that the job you are working on will be any better than what they received before.

A contractor I know gives his crew $100 when they have to spend the day in the some deep dark recess of a house sealing ductwork or installing insulation. He says that he considers places like that "Hell" and if the worker has to spend the day in "Hell" he is going to reward them for going there. The first time you do something like that, it's a surprise and the worker's level of enthusiasm is high. But they may become accustomed to it, expect it, and it may reach a level of negative feedback. Do actors get accustomed to receiving applause after a performance? If they don't get it, does the quality of their actions get worse the next time? Or do they strive to get better - to get back to the positive feedback? As the QCI you need to know how to get the crew to strive to get better with balanced feedback.

Quality work generates quality work. A satisfied customer is the best feedback a crewmember can receive.

Domain V: Ensuring Program Compliance

Task V.1 Maintain Professional Credentials

If you have gone through all the effort to achieve certification, you need to treat it as the valuable asset that it is. One of the truly beautiful elements of this business is that it is impossible to stop finding things that you don't know, things that you need to know more about. Every time you open a door there is a new world to discover that will enhance your skills and help you to do a better job every single day. Really smart people are developing new insulation, new combustion equipment, new diagnostic tools day after day and week after week. There are web sites and conferences and webinars and seminars and courses that you can take. (Upcoming conference listings: http://bit.ly/1wLsPUP) Programs like BPI and RESNET require you to keep learning, and that is a very good thing. On-line communities like Home Energy Pros (http://homeenergypros.lbl.gov/) and LinkedIn have incredible and dedicated people who are willing and excited to answer your questions and learn from you.

BPI has a certification renewal policy (http://bit.ly/1QsXkoQ). As of this writing, there are over 500 courses available where you can receive Continuing Education Units (CEUs), almost 200 of which are on-line. With enough CEUs, BPI will waive the requirement to take the written test.

BPI WRITTEN EXAM REQUIREMENTS FOR RENEWALS	
Number of CEUs	Requirement
30 or more	No written exam required
10 - 29	Need to take and pass a 50 question online exam for each designation

0 - 9	Need to take and pass a 100 question online exam for the first designation and a 50 question online exam for each additional designation

Keep track of your certification dates and go to the BPI (www.BPI.org)website and sign-in as a Candidate to check your status. (Even when you have been certified you are still a 'candidate' in BPI-world.) CEUs expire in three years. The idea is to keep current and learning more.

From the BPI QCI Scheme Handbook Revised 12/02/2014 :

The certified Quality Control Inspector will be required to renew their certification every three years. The candidate will receive an email notification at 120, 90, 60 and 30 days reminding them their certification will expire. Candidates will be allowed to start the recertification process six (6) months prior to expiration. If recertification is completed more than six (6) months in advance the expiration date will change to the date of last exam.

In order to be eligible for recertification candidates must have accumulated a minimum of 10 qualifying continuing education units (CEUs) per year over the three years of certification. When the minimum CEU requirement is met the candidate must successfully complete the field examination that is current at the time of renewal. Candidates who do not meet the CEU requirements must retake the written and field exam that is current at the time of renewal. Candidates must also re-attest to and sign the Code of Ethics to be eligible for recertification.

As part of the recertification process BPI will review the certified Quality Control Inspector's file for any open complaints.

The recertification requirements for certified Quality Control Inspectors must be completed prior to re- issuance of the certification. Should there be any open complaints at the time of expiration BPI will not award recertification.

The certification of the individual shall be withdrawn or revoked due to the certified person's negligent refusal to follow the certification scheme requirements.

Task V.2 Confirm the Allocation of Funds

There is a section of the JTA that is entitled "Allocation of Funds". Not many of us (particularly in this business) have access to unlimited budgets where we can spend anything we want! How do you evaluate what you should spend the available money on? The big issues should have been resolved long before the QCI gets out to the house and starts looking over the work-order and comparing it to the actual project. The first consideration for any on-the-job decision has to be the health and safety of the occupants of the home. There are always those times when one considers pulling out a personal wallet to pay for something that is simply not in the program's budget! Programs have funds allocated to different aspects of the project. If the oven is in horrible shape and needs to be replaced and the program is requiring that you install a mechanical ventilation system and there aren't funds for both, how do you decide which to do?

The technique of evaluating options using the Savings to Investment Ratio (SIR) where you compare what you're spending to what it will save the homeowner in the long-run does not consider the health and safety of the activity. Only the dollars.

SIR Step 1

Determine the total cost of the project. For example, if you want to replace 10 windows in the house with energy-efficient windows and energy-efficient windows cost $600 a piece, the total cost of the project would be (10 x $600) = $6,000.

SIR Step 2

Determine the useful life of the asset. The useful life is the estimated period of time an asset will be usable and able to provide a service. Continuing with the same example, assume that the useful life of the energy-efficient windows is twenty years.

SIR Step 3

Determine the annual increase (or decrease) in savings associated with the project. Continuing with the same example, assume that you would save $60 a year in energy costs by replacing 10 of your existing windows with energy-efficient windows. (Note that this only considers the conductive heat loss and not the convective heat loss).

SIR Step 4

Multiply the useful life of the product by the savings per year associated with the project. Then divide the result by the total cost of the project. Continuing with the same example, you would calculate as follows: (20 years x $60)/$6,000 = 0.20.

This is effectively saying that for every dollar you spend on the windows you would be getting twenty cents back. Most programs are looking for an SIR of at least 1 so that for every dollar spent, a dollar is returned.

Task V.3 Evaluate Installed Measures Against Criteria

How do you know that the work that has been done has met an expected level of quality? Quality workmanship is based on experience. If you have never seen a quality installation of a dual quasi-pod (if such a thing existed), how would you know if it had been installed properly? How would you know if it was upside down or inside out or backward? You might be able to make assumptions based on what you might have experienced with a similar object. And you might make some assumptions based on your parallel experience. But of course you could be completely wrong!

What you are seeing as a certified QCI, however, should be quite familiar. You should have experienced what the quality installation of loose fill cellulose insulation in an attic should look like not only because you have inspected a significant number of jobs, but also because you have actually installed loose fill installation in an attic and you know where the challenges are. Added to that are your life experiences. Some people want their driveways shoveled with perfectly sheared edges to the snow. For them, that is a quality job. Other people might consider such behavior a level of Teutonic neatness unnecessary and a waste of time.

So before you can determine if the work that has been done meets acceptable practices, you need to know what the acceptable practices are. Does the work simply need to MEET the expectations of the job or should it EXCEED those expectations? Who defines if work is "acceptable"? Getting a job done completely, on time and on budget is the most basic prerequisite. But then there is your own level of acceptable quality. And have you communicated that effectively to the crew leader and the crew? Is an excess of foam insulation, oozing out of a joint in the basement acceptable? It clearly shows that the work has been done. It's easy to see and easy to verify. It is simple to have a list of tasks and check them off: blew in fifteen inches of cellulose in the attic - check; air sealed with flashing and high temperature caulk around the chimney - check.

The work order lists adding enough insulation in the attic to average R-49. You stick your head up through the attic hatch and note that loose fill cellulose has apparently been blown-in. How can you verify that R-49 has been installed?

Original energy audit says that there were 6" fiberglass batts. Referring to the Building Analyst Standard, 6" fiberglass in Fair condition would be equivalent to R-11. If the BA standard says that loose fill fiberglass is 3.7 per inch, there would need to be about 11 inches of loose fill blown in to get to R-49.

But the work order says that the attic should average R-49. What about the attic hatch? Do the math:

If the total attic area is 600 square feet and the total hatch area is 8 square feet, the net attic area is 592 square feet. The attic is now R-49, but the hatch is only R-3.

Average R value = Gross area of the attic/((Net area of the attic/Attic R value) + (Area of the hatch/Hatch R value))

$$\text{Average R value} = 600/((592/49) + (8/3)) = 40.7$$

If the attic hatch was only R-1, the average R-value of the ceiling would be just 30! So insulating the hatch makes a massive difference in the average R-value of the ceiling. You're going to need to bring the R-value of the hatch up to 49 to average R-49 across the attic. Or the installers should have installed 12 inches of cellulose and then 2 inches of polyisocyanurate would do the trick on the hatch. You can put this into a quick spreadsheet or use the Residential Energy Dynamics tool.

This is just an example of not making assumptions and it also leads to the question of what would you do? What would you do if you did stick your head up into the attic and found an excellent insulation of 11 inches of cellulose and when you do the math you realize it's only R-41 instead of R-49? Would you get the crew to add more insulation to the hatch? Have the insulation contractor come back and blow in another inch? Or just close the hatch and sign off on the job? Did the crew meet the requirements of the work order?

————————————————————

BPI QCI Field Test

- Verified installed components listed in the work scope
- Compared work completed with acceptable practices
- Made appropriate conclusion based on the findings

- Identified work that does not meet acceptable practices
- Determined if the problem was a material problem or a work practices problem

---

Task V.4 Close Out the Project

If you started a punch-list or if you have one from your organization to close out the project, make sure that everything has been checked off.

The best documentation of work completed is photographs. Make sure that you have a means of identifying what house you are photographing. It is a pain when you have a couple of dozen pictures of details of various houses and you can't identify what houses they are! Not much use. The location tag on most cell phone images will give you an idea of what town you're in, but it won't id what house you're in. Take a shot of the work order with the address on it before you start documenting a project. You could use the house number, but you'll have to remember when you first come into the house and it might not be in a photographable spot.

Are there any certificates required? It is not code required that you attach a permanent record of what has been done, but for new houses a certificate of compliance is required to be affixed to the electrical panel. The ICC requires a certificate similar to the following:

"The certificate shall list the:

- Predominant R-values of insulation;
- U-factors for fenestration;
- The solar heat gain coefficient (SHGC) of fenestration;
- Types and efficiency of heating, cooling and service water heating equipment.

For the 2009 and subsequent editions of the IECC and IRC, the certificate shall indicate if any of the following are installed but shall not list the efficiency:

- Gas-fired unvented room heater;
- Electric furnace;
- Baseboard electric heater.

For the 2012 and subsequent editions of the IECC and IRC, the certificate shall indicate the results from any required duct system and building envelope air leakage testing done on the building."

Insulation installation certificate: Many insulation manufacturers will supply installation certificates that can be used to certify the quality, R-value, and amount of insulation installed. Most building departments require that these certificates be displayed in an obvious place in a permanent manner.

CAZ testing certification: A copy of the combustion safety testing should be included in the package for the project. Ideally, a tag or sticker should be attached to the furnace, boiler, or water heater that states when the system was tested and the results. This makes a visible and comparable document that future testing can be compared to.

Duct and Blower Door Testing: Any duct testing or blower door testing certificates should include identification of the credentials of the tester such as the BPI or HERS Rater #. It should also include what equipment was used and when it was calibrated.

Signatures: Programs require a variety of people to sign-off on a completed project. The cast of necessary characters should be listed on your punch-list.

Completion reports: These will depend on what is required by your program or organization. The file on the project should include all of the checklists and other documentation so that it can be verified.

Authorization for payments/reimbursements/invoices: And finally there are all the money issues! It is worthwhile to have a standard process for billing and payments so that elements are not missed. Bills from contractors need to be checked against the work that has been done. All those little slips of paper that come from 'gofer' runs to the hardware store, coffee, lunch, office supplies, etc. can be scanned with a portable scanner. Many companies have software that will assemble all the receipts into an organized report.

BPI QCI Field Test

- Ensured punch list items were completed
- Ensured all relevant signatures were obtained

Task V.5 Maintain Files and Records

What and how you maintain files will be up to your organization. For the weatherization agencies, you should refer to the WAPTAC (http://bit.ly/1K2Er9B)website. There are numerous Customer Resource Management programs that provide all sorts of ways of maintain information about customers and clients. You will need to be familiar with whatever system your organization requires. Familiarity with the DOE Guidelines is an integral part of the QCI testing and you should be familiar with the contents of 10 CFR 440 (http://bit.ly/1PzJwFH)and the Quality Work Plan requirements. Note that the questions on the QCI written test were formulated in 2014.

Although these are not medical records, there are certainly privacy considerations in maintaining files on houses and people. Make sure that you are familiar with what your organization and location requires, and that compliance with these standards are maintained.

United States Government
Department of Energy
Memorandum

DATE: April 8, 2015 WAP Memorandum 010

REPLY TO ATTN OF: Robert C. Adams, WAP Manager Weatherization and Intergovernmental Program Office

SUBJECT: Quality Management Plan – Recordkeeping and Reporting

TO: Weatherization Assistance Program (WAP) Grantees 10 CFR 440.24 requires, among other specific record keeping requirements, Grantees and Subgrantees administering the Weatherization Assistance Program (WAP) keep records as the Department of Energy (DOE) deems necessary for an effective audit and performance evaluation.

Based on this requirement DOE has required Grantees to keep records related to client file documentation. DOE has always accepted the ability of Grantees and Subgrantees to meet DOE's record keeping requirements related to client file documentation. However, in recent years there have been less clear and consistent approaches being used as evidenced by state and federal monitoring and oversight.

The Quality Management Plan (QMP) working group, convened by DOE provided information in order for WAP to develop a draft framework to assist Grantees and Subgrantees in how to keep records consistently and how to provide access to the documentation supporting a weatherized unit.

Regardless of the format of records, Grantee and Subgrantee systems must demonstrate:

· Program rules are being followed (e.g., eligibility requirements are being followed).

· Appropriate cost effective measures are being installed (e.g., audit results are consistent with the work order generated and the invoice costs are consistent with those estimated in the audit).

· Health and safety issues are treated according to guidance (e.g., Certified Renovator is assigned to lead-paint jobs).

· Inspections are occurring as required (e.g., 100% of the units are inspected and the post-inspection checklist includes the inspection of the audit assessment).

By establishing consistent records, Grantees and Subgrantees can continue to execute DOE's expectation of high quality management and proper documentation of WAP resources. The attached draft Client File Checklist identifies elements anticipated to be in a "complete" WAP client file. The specific contents and how the information is organized can still vary by Grantee. This is not intended to be a prescriptive list of forms or a mandated filing system. It is a tool to assist Grantees and Subgrantees in aligning file requirements to create a comprehensive client file that is able to adequately document how Weatherization services were provided in a specific home.

Link to Client (http://bit.ly/1GOwaoT)

Link to Weatherization Program Notice WPN 15-4 Quality Work Plan Requirement Update (http://bit.ly/1FjtF8B)

BPI QCI Field Test

- Discussed how job logs and notes are maintained
- Discussed how information on active complaints is maintained
- Discussed how information on job anomalies is maintained

Supporting Documents and Links

Taking Multiple Choice Tests

In general the role of QCI requires a lot of documentation of situations. Answers that include the word 'document' are likely to be the right answer.

To prepare for a multiple-choice exam, consider the following steps:

Begin studying early: *The QCI, multiple-choice exam focuses on details, and you cannot retain many details effectively in short-term memory. If you learn a little bit each day and allow plenty of time for repeated reviews, you will build a much more reliable long-term memory.*

Make sure that you identify and understand thoroughly both the technical and administrative issues. *Pay particular attention to fundamental terms and concepts that describe important events or features, or that tie related ideas and observations together. These are the items that most commonly appear on multiple-choice exams.*

As you study your class notes and your assigned readings, make lists and tables. *Concentrate on understanding multi-step processes, problem solving, and on ideas, events, or objects that form natural sequences or groupings. Look for similarities and differences that might be used to distinguish correct choices from distracters on an exam.*

Be sure you understand new vocabulary and key definitions. Check to be sure that you have not left any out by mistake.

Do not simply memorize the book's definitions. You must be sure that you really know what the definitions mean.

Brainstorm possible questions with several other students.

Practice on sample questions so that you understand the flavor of the QCI exam.

Answering Multiple Choice Questions There are many strategies for maximizing your success on multiple choice exams. The best way to improve your chances, of course, is to study carefully before the exam. There is no good substitute for knowing the right answer. Even a well-prepared student can make silly mistakes on a multiple choice exam, however, or can fall prey to distracters that look very similar to the correct answer.

Here are a few tips to help reduce these perils:

Remember that quality control is often dependent on documenting a situation. If the word document or note is in the answer, it may be the correct one.

If you see the response that you anticipated, note it and then check to be sure that none of the other responses is better. If you do not see a response that you expected, then consider some of the following strategies to eliminate responses that are probably wrong.

None of these strategies is infallible although they have been helpful for some people. A smart exam developer will avoid writing questions for which these strategies work, but you can always hope for a lapse of attention.

Responses that use absolute words, such as "always" or "never" are less likely to be correct than ones that use conditional words like "usually" or "probably."

"Funny" responses are usually wrong.

"All of the above" is often a correct response. If you can verify that more than one of the other responses is probably correct, then choose "all of the above."

"None of the above" is usually an incorrect response, but this is less reliable than the "all of the above" rule. Be very careful not to be trapped by double negatives.

Look for grammatical clues. If the stem ends with the indefinite article "an," for example, then the correct response probably begins with a vowel.

The longest response is often the correct one, because the question designer tends to load it with qualifying adjectives or phrases.

Look for verbal associations. A response that repeats key words that are in the stem sentence is likely to be correct.

If all else fails, choose response (b) or (c). Many question designers subconsciously feel that the correct answer is "hidden" better if it is surrounded by distracters. Response (a) is usually least likely to be the correct one.

If you cannot answer a question within a minute or less, skip it and plan to come back later.

Take the time to check your work.

Choose one of the answers even if you don't have a clue which one is right. A blank is definitely the wrong answer.

Source: Center for Teaching Excellence

What To Do If More Than One Answer Seems Correct

If you're utterly stumped by a question, here are some strategies to help you narrow the field and select the correct answer:

Ask yourself whether the answer you're considering completely addresses the question. If the test answer is only partly true or is true only under certain narrow conditions, then it's probably not the right answer. If you have to make a significant assumption in order for the answer to be true, ask yourself whether this assumption is obvious enough that the question designer would expect everyone to make it. If not, dump the answer overboard.

If you think an item is a trick question, think again. Very few question designers would ever write a question intended to be deceptive. If you suspect that a question is a trick item, make sure you're not reading too much into the question, and try to avoid imagining detailed scenarios in which the answer *could* be true. In most cases, "trick questions" are only tricky because they're not taken at face value.

If, after your very best effort, you cannot choose between two alternatives, try vividly imagining each one as the correct answer. If you are like most people, you will often "feel" that one of the answers is wrong. Trust this feeling -- research suggests that feelings are frequently accessible even when recall is poor (e.g., we can still know how we feel about a person even if we can't remember the person's name). Although this tip is not infallible, many students find it useful.

Based on info from: https://www.socialpsychology.org/testtips.htm

Test Taking Guidance from CHP (Community Housing Partners):

HEP (Home Energy Professional) BPI QCI Testing Reminders

Field Exams:

- The video camera is not to video you, but to listen to us as proctors;
- Use your field guide – make sure you talk about or demonstrate each item fully;
- If something doesn't look "right" (manometer, combustion analyzer etc..) check settings and if needed start over – this way you don't miss a step or take the wrong readings;
- Consult correct BPI standards and make recommendations based on charts & action tables;
- Read fully the sample audit and work scope;
- Make notes about the house
- Make recommendations against diagnostic tests – compare to pre-and post-readings
- Keep in mind that the audit and work scope may or may not be accurate....
- You can use all notes and anything that will help you during the test – plan accordingly
- Written Exam:
    - Use your scratch paper to draw out house scenarios
    - Use Formula sheet – if you want a printed one ask
    - Work calculations based on formula sheet
- Read the BPI Building Analyst standards before test and consult during test
- Read questions thoroughly – pay attention to key words: any answer with EXCEPT; NOT; AND, WOULD NOT, is the best option; ...
- READ ALL ANSWERS FIRST, then eliminate 2 obvious wrong answers then select the 1 that will work 100% of time
- DO NOT apply your agency or state protocols to the questions, keep in mind that some answers may be right even though you do not do these in your state
- DO think from the perspective of this being a NATIONAL EXAM
- Apply the concept of you being an in-house QCI inspector unless otherwise stated.

———————

Don't Psych Yourself Out

It's been demonstrated that when you carry extra emotional baggage -- "I've got to ace this exam" or "If I screw up, I'll lose my job" -- performance suffers, so don't lose the big picture. The most constructive approach is to focus on the task at hand, put in as much time studying as you can afford, and just do your best. Returning to the luggage metaphor: all the excess baggage can be dealt with later, if you're so inclined. Unpack *after* the exam.

A certain amount of anxiety is normal (or even useful) when studying for an exam, but if you feel overwhelmed or feel that uncontrollable emotions are interfering with your exam performance, you may be suffering from test anxiety. If you think this is a possibility, you should alert your instructor right away and ask for help. Your instructor may be able to recommend techniques to reduce your anxiety (e.g., relaxation training) or may allow you to take the exam in a less stressful environment (e.g., by moving you to a quieter room, or giving you a few extra minutes to complete the exam). In a case of severe test anxiety, the worst thing you can do is to continue taking the exams without notifying your instructor that there is a problem.

——————

Finally, Attitude again

It's not the most unique observation, but what do they call a pilot who finishes at the bottom of the class? A pilot. There is a lot here. It takes a lot to be a Quality Control Inspector, but you wouldn't be reading this or challenging the QCI exams if you weren't capable of passing them . . . if you didn't have the ATTITUDE to get all the way to the top of the mountain. It's a matter of knowing the tools, understanding what they are telling you and being able to think through logical, decision-making. I wish you the best of luck on your exams and on your career as a Certified Quality Control Inspector!

End Notes etc.

Ventilation system testing protocols

These are standardized testing protocols for various residential ventilation systems. They are designed to test the airflow through the ventilation system and not verify the performance of the controls or the effectiveness of the overall design.

Exhaust only systems

The primary purpose for a large majority of exhaust systems has been to ventilate a bathroom or kitchen, serving as a local exhaust system, removing pollutants close to the source. These systems can also be used for whole building ventilation. When they are used for whole building ventilation, they need to have clear access to the rest of the volume in the house. But when they are used for local exhaust, it is common (particularly in a bathroom) to keep the bathroom door closed.

Testing bathroom-type or single port or multi-port in-line fans for whole house ventilation:

1. Open all interior doors in the house;
2. Doors and windows to the exterior must be closed;
3. Fireplace dampers must be closed;
4. The air handlers and all exhaust fans except the unit under test must be turned off;
5. Bathroom or kitchen door must be open;
6. Measure the flow with the appropriate tool.

Testing bathroom-type or single port or multi-port in-line fans for local exhaust ventilation:

1. Doors and windows to the exterior in the room being tested must be closed;
2. Bathroom or kitchen door must be closed;
3. Measure the flow with the appropriate tool.

Supply-only systems

Supply-only systems pressurize the house. They are applicable to cooling dominated climates. These systems put the house under positive pressure and force the air out through cracks in the structure. A low volume of supply air is provided to spaces where people spend the most time such as bedrooms.

Testing an in-line supply-only fan for whole house ventilation:

1. Open all interior doors in the house;
2. Doors and windows to the exterior must be closed;
3. Fireplace dampers must be closed;
4. The air handlers and all exhaust fans except the unit under test must be turned off;
5. Prepare a hole in the supply duct on the low pressure side of the fan;
6. Using a pitot tube or averaging flow sensor inserted into the intake duct on the low pressure side of the fan, measure the pressure and calculate the airflow.

An alternative (same house setup as above):

1. Prepare a hole in the duct on the low pressure side of the supply fan;
2. Turn on the supply fan;
3. Insert a static pressure probe into the duct and measure and record the static pressure;
4. Disconnect the outside air duct where it connects to the fan;
5. Connect the outside air duct to a calibrated fan such as a duct tester so that the duct is depressurized;
6. Use the duct tester to bring the pressure in the outside air duct back to the same pressure measured in Step 3;
7. Record the measured flow through the duct tester. This is equivalent to the flow through the supply fan and would equal the whole building supply ventilation;
8. Restore the connections when the measurement is complete.

An outside air connection to the return side of an air handler can be used to provide fresh air through the air handler when the air handler blower is running. Some of these systems include a control that will cycle the air handler on periodically to meet the ventilation requirement. Some of the systems also include a motorized damper to keep the outside air supply closed when ventilation air is not required.

Testing a supply-only system connected to the return side of an air handler:

1. Open all interior doors in the house;
2. Doors and windows to the exterior must be closed;
3. Fireplace dampers must be closed;

4. Motorized dampers in the outdoor air supply duct must be open;
5. Prepare a hole in the outside air supply duct on the low pressure side of the fan;
6. Turn on the air handler and using a pitot tube or averaging flow sensor inserted into the intake duct on the low-pressure side of the fan, measure the pressure and calculate the airflow.

An alternative (same house setup as above):

1. Prepare a hole in the outside air supply duct on the low pressure side of the fan;
2. Turn on the air handler;
3. Insert a static pressure probe into the duct and measure and record the static pressure;
4. Disconnect the outside air duct where it connects to the return duct;
5. Connect the outside air duct to a calibrated fan such as a duct tester so that the duct is depressurized;
6. Use the duct tester to bring the pressure in the outside air duct back to the same pressure measured in Step 3;
7. Record the measured flow through the duct tester. This is equivalent to the flow through the supply fan.
8. Restore the connections when the measurement is complete.

Balanced systems

Balanced systems are designed to bring the same amount of air into the house as they extract from the house. The two air streams are in proximity to each other as they move through an exchanger core. The streams are separated by the material of the core and never actually touch each other, but the large surface areas of the core allows a significant amount of heat (heat recovery ventilators (HRV)) or heat and moisture (enthalpy or energy recovery ventilators (ERV)) to be transferred from one stream to the other. Their transfer efficiency is established in a laboratory where the flow through the two steams is kept balanced. If the flows are not balanced, the efficiency of the system will change.

These systems will have multiple supply and exhaust points. Note that flow-measuring products such as flow hoods are not effective at measuring low supply airflows.

Testing the flow through a stand-alone HRV or ERV (not connected to either side of an HVAC system):

1. HRV or ERV must be fully installed and connected;
2. Open all interior doors in the house;
3. Doors and windows to the exterior must be closed;
4. Fireplace dampers must be closed;
5. Filters and core of the ventilator must be clean;
6. Supply and exhaust termination fittings on the outside of the house must be clear and open;
7. HRV clear of ice and drain pan not full of water;

8. Install flow measuring stations[7] in the supply and exhaust connections from and to the outside (or use the flow taps on the ventilator unit provided by the manufacturer);
9. Turn the ventilator on high speed;
10. Measure the velocity pressure with a single channel of the manometer connected to the two ports on the flow station[8];
11. Convert the reading to cfm using the calibration label attached to the flow station or by using the conversion formula $V = 1096.2 \times (V_p/0.075)^{0.5}$ where V equals the velocity in feet per minute and V_p is the velocity pressure in inches of water gauge;
12. Multiply the V_p by the area of the ducting;
13. Repeat the process for the other air stream. The two flows should be within 10% of each other. The average flow of the two air streams will be the ventilation rate through the system.

Testing the flow through an HRV or ERV connected to either side of an HVAC system. This will require two tests: one with the air handler turned off and one with the air handler turned on. If the ventilator only runs when the air handler runs, then only Test 1 needs to be performed:

Test 1 Air Handler turned on:

1. HRV or ERV must be fully installed and connected;
2. Open all interior doors in the house;
3. Doors and windows to the exterior must be closed;
4. Fireplace dampers must be closed;
5. Filters and core of the ventilator must be clean;
6. Filter in the air handler must be clean;
7. Supply and exhaust termination fittings on the outside of the house must be clear and open;
8. HRV clear of ice and drain pan not full of water;
9. Install flow measuring stations in the supply and exhaust connections from and to the outside (or use the flow taps on the ventilator unit provided by the manufacturer);
10. Turn the ventilator on high speed;
11. Turn the air handler fan on;
12. Measure the velocity pressure with a single channel of the manometer connected to the two ports on the flow station;

---

[7] Air turbulence can have a significant effect on the measurements, so the flow measuring station should be located far enough away from turbulence producing elements (like elbows, dampers, or the fans themselves to get accurate readings. As a rule of thumb, the flow measuring station should be located a minimum of 12 inches (300 mm) away from dampers and 30 inches away from axial (propeller) fans or the outlets of blowers.

[8] Connecting both taps on a single channel of the manometer will provide the difference between the input and the reference taps ie the difference between the Total Pressure and the Static Pressure which is the Velocity Pressure.

13. Convert the reading to cfm using the calibration label attached to the flow station or by using the conversion formula $V = 1096.2 \times (V_p/0.075)^{0.5}$ where V equals the velocity in feet per minute and V_p is the velocity pressure in inches of water gauge;
14. Multiply the V_p by the area of the ducting;
15. Repeat the process for the other air stream. The two flows should be within 10% of each other. Note the average flow of the two air streams. This is the whole building ventilation rate if the ventilator only runs when the air handler is running.
16. If the ventilator is operated on an independent control continue with Test 2.

Test 2 Air Handler turned off:

1. Air handler fan must be turned off
2. Repeat steps 2 – 10 above;
3. Turn the ventilator on high speed;
4. Measure the velocity pressure with a single channel of the manometer connected to the two ports on the flow station;
5. Convert the reading to cfm using the calibration label attached to the flow station or by using the conversion formula $V = 1096.2 \times (V_p/0.075)^{0.5}$ where V equals the velocity in feet per minute and V_p is the velocity pressure in inches of water gauge;
6. Multiply the V_p by the area of the ducting;
7. Repeat the process for the other air stream. The two flows should be within 10% of each other. Note the average flow of the two air steams;
8. The average of the two tests (with the air handler on and the air handler off) is the whole building ventilation rate.

Flow measuring tools

Note that the accuracy of airflow measuring tools is typically about ± 10%.

- TEC EXH
    - DG700
    - DM32
- FlowFinder2
- Flow Blaster
- Testo Large Vane Anemometer
- TSI Velocicalc
- Pitot tube (Averaging Flow Sensor PAFS (Dwyer) or Kele Differential Pressure Flow Sensor DPFS)

Using an Averaging Flow Sensor (AFS) to measure airflow

An averaging flow sensor has taps for two pressure hoses. The upstream port (H) senses *total pressure.* The downstream port (L) senses *static pressure.* The difference between these is the *velocity pressure.* The sensor inside the duct has openings on both the upstream and the downstream sides. The velocity pressure can be used to determine the airflow velocity in feet per minute (fpm) and converted to provide airflow volume (cfm). The square root of the velocity pressure times the constant (253.29) equals the velocity of the airflow in feet per minute and multiplying that by the area of the duct provides the cubic feet per minute (cfm).

Note: Refer to http://bit.ly/1zFY0Pk 4" Sensor K factor = 1.32, 6" sensor K factor = 1.39

Installation:

If possible, install the unit horizontally to assure accurate velocity reading;

Install in a duct section that will provide a smooth flow of air – at least 2x duct diameters downstream from a hard elbow (8" for a 4" AFS, 12" for a 6" AFS);

Drill a 7/8" hole in the duct;

Attach using (2) self-tapping screws inserted into the 3/16" mounting holes;

Attach the H port to the Input on the manometer;

Attach the L port to the Reference on the same channel on the manometer;

Reading will be the VP (velocity pressure).

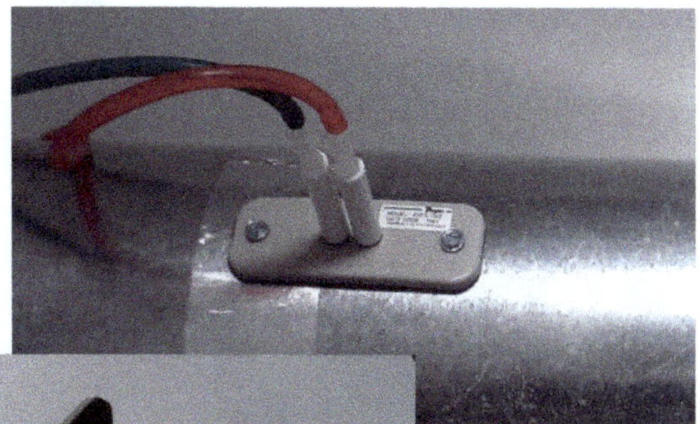

Air movement through a duct measured with an averaging flow sensor:

$$VP = TP - SP \ (\Delta P \text{ in Pascals})$$

$$FPM = VP \times 253.29$$

$$CFM = A \times FPM$$

Where:

VP = velocity pressure in Pascals

TP = total pressure in Pascals

SP = static pressure in Pascals

FPM = feet per minute

CFM = cubic feet per minute

A = area of the duct in square feet

4" diameter duct = 0.087 ft^2

6" diameter duct = 0.196 ft^2

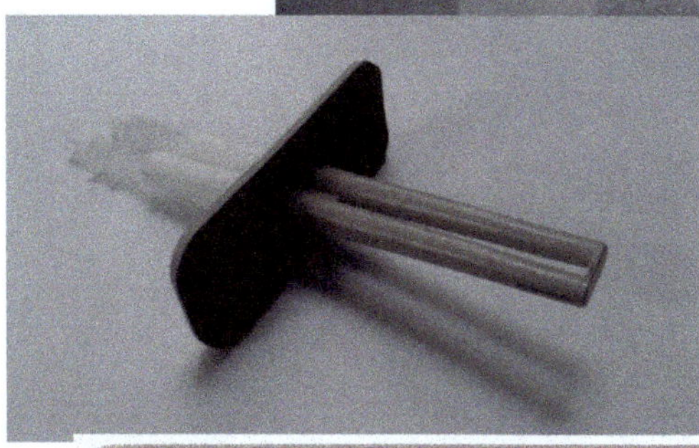

Do the math or plug the information into RED!

Example (4" duct):

Measured velocity pressure (VP) = 1.85 Pascals

FPM = $(1.85)^{-2}$ x 253.29 = 1.361 x 253.29

FPM = 344.8

CFM = 0.087 x 344.8

CFM = 30 (Does not include pitot tube K factor.)

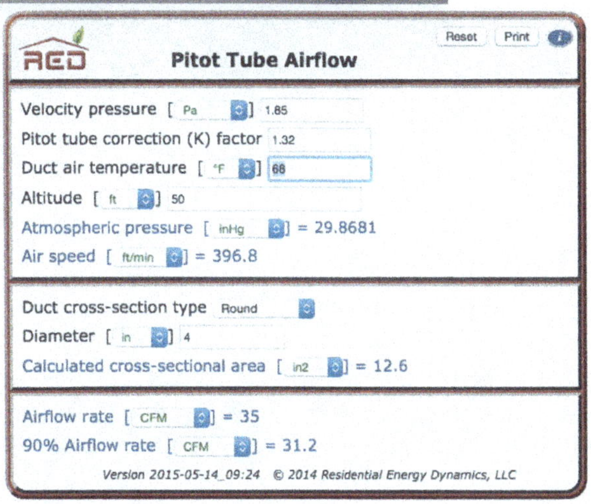

Range hood testing using a duct tester

Using duct tester as a powered flow hood for measuring the flow through a range hood is an accurate way to measure range hood flows in excess of 100 cfm without the difficulties of getting to the hood on the exterior of the building or dealing with local wind conditions.

1. Create a baffle to cover the opening of the range hood with an opening for the duct tester fitting.
2. Attach the baffle to range hood. (This is sometimes difficult due to grease, which won't allow the tape to stick well.) The duct tester's ducting adds weight to the baffle, trying to pull it off.
3. Attach the duct tester's ducting (snorkel) to the fitting on the baffle.

4. Connect the manometer and fan:
   a. DG700
      i. Green hose to Input Tap on Channel A. Insert the other end into the range hood cavity.
      ii. Red hose to Input Tap on Channel B. Connect the other end to the brass tap in middle of the DB fan housing.
   b. DM2A or DM32
      i. Blue hose to Input Tap on Channel A. Insert the other end into the range hood cavity.
      ii. Yellow hose to Reference Tap on Channel B. Connect the other end to the Ref B tap on the duct tester fan.
      iii. Green hose to the Input Tap on Channel B. Connect the other end to the "B" tap on the duct tester fan.

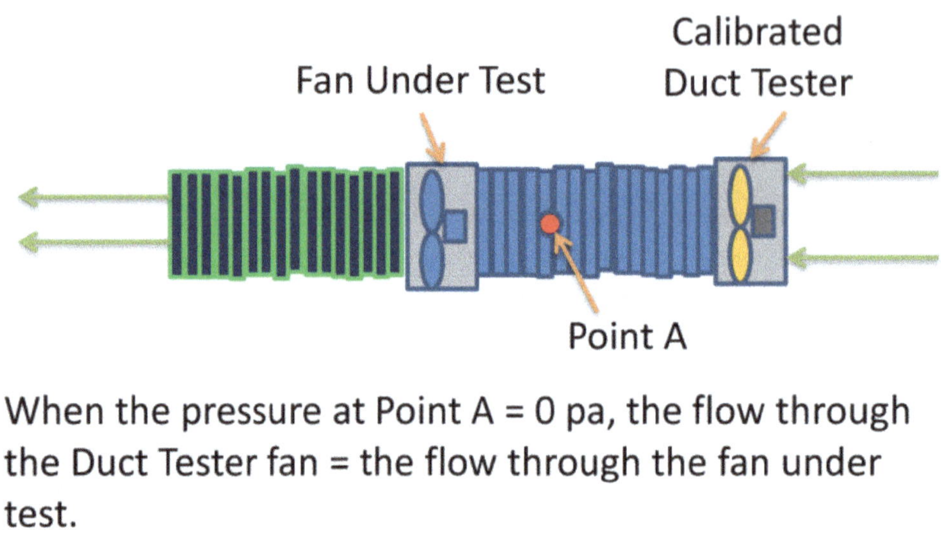

5. Turn on the range hood to full speed. Pressure reading on Channel A of the manometer should drop, moving increasingly negative as the duct of the duct tester goes negative.
6. Turn on the duct tester fan, increasing the flow until the pressure on Channel A becomes 0 Pascals. At that point the flow through the duct tester fan will be equal to the flow through the range hood fan and the flow reading can be made on Channel B.
7. Repeat the test for other fan speeds.

The same process can be used to test bathroom fans. The duct tester can be connected a box that covers the fan or grille. Because of the powered flow into the fan being tested, the measuring device (duct tester) is essentially invisible to the fan being tested, offering no resistance or impediment to the flow.

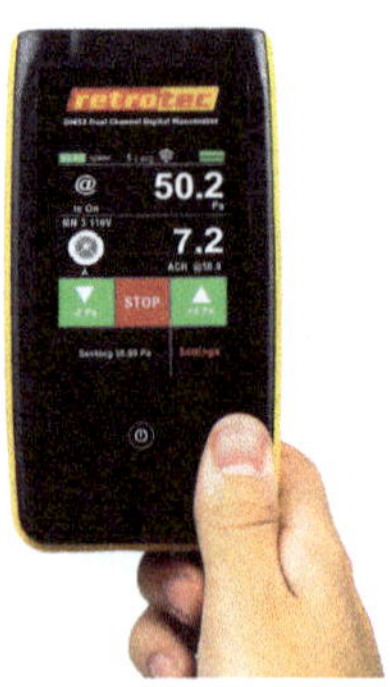

Bath fan testing using a TEC EXH

The Exhaust Fan Flow Meter (EXH) is very similar to a pressure pan, but it is deeper and includes an adjustable door or opening, and a soft, rubber gasket around the edge that will seal the EXH to the surface. On its own it doesn't tell you much. It must be connected to a manometer or pressure gauge. If you tell the manometer the size of the opening and the pressure difference across it, the manometer can calculate the airflow. The EXH is calibrated to the TEC (The Energy Conservatory) DG700 digital manometer or Retrotec DM2 or DM32 (but will work with any digital manometer). The maximum flow that can be accurately measured with the EXH is 124 cfm, which is acceptable for most bathroom fans. It comes with a Velcro pad that can be attached to a painter's pole to help you reach up and hold the EXH over a fan's ceiling grille.

The pressure is read on Channel A and the airflow on Channel B as it is for the Blower Door or Duct Tester. Since there is only a single hose connection on the EXH, the two channels on the manometer should be jumped together and attached to the single hose running to the EXH. This will allow both the pressure and the flow to be measured simultaneously. If the pressure exceeds 8 Pascals, the readings will lose accuracy.

1. Connect a jumper hose to the input taps of both manometer channels;
2. Connect the hose to the EXH;
3. Select the *MODE* setting of *Pressure/Flow*;
4. Use the *DEVICE* control to select the EXH option;
5. Adjust the opening on the EXH to E1 and use the *Config* button to select A1 on the manometer;
6. If the pressure reading on Channel A exceeds 8 Pascals, open the door wider and select the corresponding setting on the manometer.

You can use the EXH with other manometers by reading the pressure across the opening and referring to the table attached to the side of the EXH housing. The numbers on the chart correspond to the manometer's pressure readings.

A C System Basics

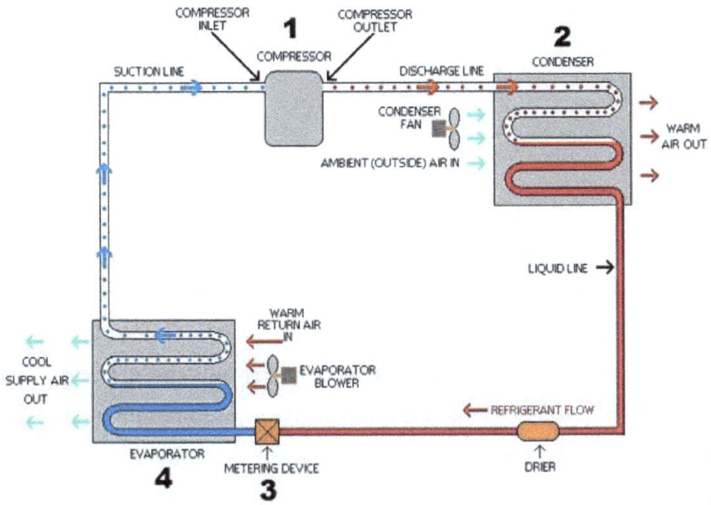

Compressor: The component at #1 in this air conditioning circuit and cycle diagram is the compressor. The compressor is the heart of the system; it keeps the refrigerant flowing through the system at specific rates of flow, and at specific pressures. It takes refrigerant vapor in from the low-pressure side of the circuit, and discharges it at a much higher pressure into the high side of the circuit. The rate of flow through the system will depend on the size of the unit, and the operating pressures will depend on the refrigerant being used and the desired evaporator temperature.

Condenser: The component at #2 in this air conditioning circuit and cycle diagram is the condenser. The red dots inside the piping represent discharge vapor. The solid red color represents high-pressure liquid refrigerant. Most air-cooled air conditioning and refrigeration systems are designed so that the refrigerant will condense at a temperature about 25 to 30 degrees above outside ambient air temperature. When the hot refrigerant vapor discharged from the compressor travels through the condenser, the cool air flowing through the condenser coil absorbs enough heat from the vapor to cause it to condense. If the outside air temperature is 80 degrees, the system is designed so that the temperature of the refrigerant, right at the point where it first condenses, will be about 105 to 115 degrees.

The refrigerant needs to condense at this relatively high temperature so that the air will be very cold relative to the temperature of the discharge vapor, which will allow the latent heat energy in the vapor to transfer over to that relatively cold air, and cause the refrigerant to condense. This transfer of heat from the vapor to the flowing air is what makes hot air blow out of your air conditioner's condensing unit. At this stage in the air conditioning circuit and cycle diagram, high pressure liquid refrigerant will flow down the liquid line, through a filter drier that is designed to prevent contaminants from flowing through the system, and on to the metering device.

Metering Device: The metering device, component #3 on this air conditioning circuit and cycle diagram, is the dividing point between the high pressure and low pressure sides of the system, and is designed to maintain a specific rate of flow of refrigerant into the low side of the system. If the wrong capacity of metering device is used, or if there is a problem with the metering device, an incorrect quantity of refrigerant will flow into the evaporator. When the refrigerant passes through the metering device, it drops from about 225 psi to about 70 psi, it also drops in temperature from about 110 degrees to about 40 degrees, it starts evaporating immediately, and it wouldn't be too inaccurate to imagine it acting like warm soda when you shake the bottle and pop the top off. It shoots out into the evaporator foaming, bubbling, and boiling, and remember: it's at a low pressure, so it's only boiling at about 40 degrees F. (Lower efficiency HVAC air conditioner and heat pump models have fixed orifice types while higher efficiency systems have thermostatic expansion valves or TXV installed in the HVAC air conditioner or heat pump system. The **thermostatic expansion valve** is far more complex than the fixed orifice metering device as the thermostatic expansion valve metering device modulates the refrigerant flow based on the temperature of the refrigerant temperature in the evaporator coil.)

Note that the unit may be capable of employing a TXV metering device but that does not mean that it has been installed. It should be visually verified prior to testing. A Superheat test can be used if there is not a TXV metering device. A subcooling test should be used if there is a TXV metering device.

Evaporator: And that brings us to the evaporator, component #4 in the air conditioning circuit and cycle diagram. There will be relatively warm air flowing over the evaporator coil, - about 80 degrees. The air conditioning system is designed so that the refrigerant will evaporate in the evaporator at a temperature of about 40 degrees, so that it will be cold compared to the warm air flowing over it. The system is designed so that the heat in the warm air flowing over the evaporator will be absorbed by the cold evaporating refrigerant. This cools the air flowing over the evaporator, and is the reason cold air blows out of your air conditioner.

Air-Conditioning-and-Refrigeration-Guide.com

A note about Heat pumps

· Heat pumps don't generate heat. The heat already exists.

· They are exactly like an air conditioner- they move the heat from one place to another.

· The only difference is that in heat pumps there is a Reversing Valve (http://bit.ly/1EL49s5)that allows the system to move the heat from inside the house to the outdoors (cooling mode), or to reverse the cycle and remove the heat from outside of the house to the indoors (heating mode).

· Air conditioners don't have a Reversing Valve so they can only move the heat from inside the house to the outdoors (cooling mode only).

· Note that a heat pump thermostat is different from a regular air conditioner thermostat.

Useful Formulas

Converting CFM_{50} to ACH_{50}:

$$ACH_{50} = (CFM_{50} \times 60)/V$$

Converting ACH_{50} to ACH_n:

$$ACH_n = ACH_{50}/n$$

Where:

ACH_{50} = Air changes at a 50 Pascals pressure difference

ACH_n = Air changes under natural conditions

V = volume of the house in cubic feet

n = the LBL correlation factor (Refer to LBL table or default to 20)

Calculating the approximate airflow through a hole when you know the area of the hole and the pressure difference:

$$Q = 1.07 \times A \times \Delta P$$

Where:

Q = airflow in cfm

A = area of the hole in square inches

ΔP = the difference in pressure in Pascals

Calculating the ASHRAE 62.2-2013 Infiltration credit:

$$Infiltration\ Credit = 0.052 \times CFM_{50} \times S_f \times W_f$$

Where:

Infiltration Credit in cfm to be subtracted from whole building ventilation requirement

W_f = weather factor from TMY3 table for specific city (from 0.29 Hilo, HI to 1.14 Barrow, AK) 0.56Av.

S_f = story factor (height of the building in stories)

1 story = 1   1.5 stories = 1.18   2 stories = 1.32   2.5 stories = 1.44   3 stories = 1.55

Calculating the static pressure in a duct run when you know the effective length of the duct and the cfm:

$$Static\ pressure\ (for\ rigid\ duct) = 0.002 \times L/D \times V/1096.52$$

$$Static\ pressure\ (for\ flex\ duct) = 0.004 \times L/D \times V/1096.52$$

Where:

Static pressure in iwg (to convert to Pascals, multiply by 248.8)

L = effective length (actual + equivalent length) of the ducting in feet
D = diameter converted to feet
V = velocity of the air movement in fpm (cfm/area of the duct in square feet)

References and Resources

I have included a number of documents here that I found particularly helpful in working through the QCI JTA. Frankly I wish I could just include the printed versions of all these things because I have found that lists like this are rarely taken advantage of and frankly some of these links and reports say things a lot better than I do. However, in the interest of saving trees, I will include them here electronically (for the most part). But I would urge you not to ignore them. You don't have to study them to get through the QCI exam, but they will help you to do a better job as a Quality Control Inspector.

Arizona Critical Details Field Guide

The Arizona Critical Detail guide is focused on Arizona but the fundamentals apply throughout the country: (Note: this document takes quite awhile to download.) AZ Critical Details (CD) Guide (http://bit.ly/1KwdzNd)

BPI Building Analyst Standard:

http://www.bpi.org/tools_downloads.aspx?selectedTypeID=1&selectedID=2

This is one you have to have. There is a pull down for it on the written exam, but you really want to be sure that you are familiar with where things are so you don't spend a bunch of time fiddling around looking for things. (BPI is planning to phase this standard out in January of 2017.)

BPI QCI Field Guide

http://www.bpi.org/files/pdf/Quality%20Control%20Field%20Guide.pdf

This is the field guide that the field test is based on. The proctor needs to see the candidate go through all of these steps.

Calibre Book Conversion Software

http://calibre-ebook.com/

This is fantastic software for the conversion of eBooks from one format to another. It's free. It works really well. And it does a lot of stuff. How can you go wrong with that?

Combustion Safety For Natural Draft Appliance Using Indoor Air:

http://www.nrel.gov/docs/fy14osti/61326.pdf

"This measure guideline covers how to assess and carry out the combustion safety procedures for appliances and heating equipment that use indoor air for combustion in low-rise residential buildings. Only appliances installed in the living space, or in an area freely communicating with the living space, vented alone or in tandem with another appliance are considered here. A separate measure guideline addresses combustion appliances located either within the living space in enclosed closets or side rooms or outside the living space in an adjacent area like an attic or garage that use outdoor air for combustion."

Combustion Safety For Natural Draft Appliances through Appliance Zone Isolation:

http://www.nrel.gov/docs/fy14osti/61295.pdf

"This measure guideline covers how to assess and carry out the isolation of natural draft combustion appliances from the conditioned space of low-rise residential buildings. It deals with combustion appliances located either within the living space in enclosed closets or side rooms or outside the living space in an adjacent area like an attic or garage. This subset of houses does not require comprehensive combustion safety tests and simplified prescriptive procedures can be used to address safety concerns. This allows residential energy retrofit contractors inexperienced in advanced combustion safety testing to effectively address combustion safety issues and allow energy retrofits."

Both of these documents and some other really interesting stuff can be found at:
http://www.osti.gov/scitech/search.jsp

Combustion Safety Tech Brief

http://www.waptac.org/data/files/Website_Docs/technical_tools/CombustionApplianceSafety-TechBrief.pdf

This is a full color brief explanation of a most of the information that you need to do combustion safety testing. It summarizes where many different types of units need to be probed to prove that they are functioning properly.

DHW: What temperature is your hot water?

http://www.oneidanation.org/uploadedFiles/What%20Temperature%20is%20your%20Hot%20Water%20Heater.pdf

This is a useful safety brochure on the injuries caused by excessive hot water temperatures.

DOE Quality Work Plan

http://www.waptac.org/Technical-Tools/DOE

This is the fundamental place to find all of the useful information about what should be included in the memory of a QCI!

CEDA Weatherization Videos

https://www.youtube.com/user/CEDAWeathization

Community and Economic Development Association. These are some great, fundamental training videos.

EERE Quality Work Plan

http://energy.gov/eere/wipo/quality-work-plan-requirements

Here's even more information about quality work.

EERE Quality Work Plan Training Requirement

http://energy.gov/eere/wipo/quality-work-plan-training-requirement

This is where you'll find the information on the Tier 1 and Tier 2 training that is required. Once again, it goes beyond just minimal requirements covering information that will be useful on the job.

High Efficiency Natural Gas Furnaces

http://www.nrel.gov/docs/fy13osti/55493.pdf

This Measure Guideline covers installation of high-efficiency gas furnaces. Topics covered include when to install a high-efficiency gas furnace as a retrofit measure, how to identify and address risks, and the steps to be used in the selection and installation process. The guideline is written for Building America practitioners and HVAC contractors and installers. It includes a compilation of information provided by manufacturers, researchers, and the Department of Energy as well as recent research results from the Partnership for Advanced Residential Retrofit (PARR) Building America team.

Lead Hazard Renovation Form

http://www2.epa.gov/sites/production/files/documents/pre-renovationform.pdf

Lead training is required for all these activities, so although it is not listed on the prerequisite below, you should have lead safety training. Ordinary renovation and maintenance activities can create toxic lead dust that can harm your customers, workers, and even yourself. But by following lead-safe work practices, you can prevent lead hazards.

PART 440—WEATHERIZATION ASSISTANCE FOR LOW-INCOME PERSONS

http://bit.ly/1HuXoSO

This is the document that the programs are built around. Appendix A lists the standards that the materials that are used in the programs need to meet.

This part implements a weatherization assistance program to increase the energy efficiency of dwellings owned or occupied by low-income persons or to provide such persons renewable energy systems or technologies, reduce their total residential expenditures, and improve their health and safety, especially low-income persons who are particularly vulnerable such as the elderly, persons with disabilities, families with children, high residential energy users, and households with high energy burden.

PART 600 – FINANCIAL ASSISTANCE RULES

http://bit.ly/1K0QXnT

This part implements the Federal Grant and Cooperative Agreement Act, Pub. L. 95-224, as amended by Pub. L. 97-258 (31 U.S.C. 6301-6308), and establishes uniform policies and procedures for the award and administration of DOE grants and cooperative agreements. This subpart (Subpart A) sets forth the general policies and procedures applicable to the award and administration of grants, cooperative agreements, and technology investment agreements. The specific guidance for technology investment agreements is contained in part 603.

Prerequisites for Quality Control Inspector Certification

All items below are required prior to taking the written and/or field certification exams.

You must obtain a minimum of **40 points** from any combination of activities below:

- Industry inspector specific experience (site visits, inspections, diagnostics, etc); **maximum of 20 points;**
  - 10 points for 1,000 hours;
- Other industry specific experience (crew leader / energy auditor) minimum 2,000 hours in either activity;
  - 5 points for crew leader;
  - 10 points for energy auditor completing a minimum of 15 audits (proof required).
- Building experience (framing, roofing, drywall, siding, etc); **maximum of 10 points**
  - 5 points for each 1,000 hours;
- Training from industry specific training center (training whose content can be matched against and lines up with the content of the job task analysis for the certification); maximum of 10 points;
  - 5 points for every 40 hours

- Industry certifications (RESNET, BPI, NATE, EPA). Other certifications also considered; **maximum of 10 points**
  - 5 points per certification

QCI Certification Scheme Handbook

http://www.bpi.org/files%5Cpdf%5CQuality%20Control%20Inspector%20Scheme%20Handbook.pdf

This certification scheme handbook outlines the knowledge, skills and abilities requirements for Quality Control Inspectors. This scheme defines a Quality Control Inspector as the following; A quality control inspector is a residential energy efficiency professional who ensures the completion, appropriateness, and quality of energy upgrade work by conducting a methodological audit/inspection of the building, performing safety and diagnostic tests, and observing the work. This document is intended to include all of the tasks a Quality Control Inspector may perform, as well as the knowledge, skills, and abilities required to do these tasks.

QCI Job Task Analysis (JTA)

http://www1.eere.energy.gov/wip/pdfs/51670.pdf

The QCI JTA was put together by a group of industry experts and covers all the Knowledge, Skills, and Abilities that could reasonably fit into the document. There are a lot more things that a Quality Control Inspector needs to know, but this is a great starting point and it is the basis for the certification.

Sample QCI Written Test

http://energy.gov/sites/prod/files/2014/11/f19/qci_pre_exam_quiz_11242014.pdf

A short sample written test. Here's the thing about sample written tests: they are samples! They aren't the real test. It is important that you get the right answers but it is much more important that you understand the right answers! These are not the real questions so just memorizing the answers will only get you in the ballpark. They won't win the game.

You Tube QCI Exam

https://www.youtube.com/watch?v=ntD64TEBK2U

Dan Hughes video of an almost 2 hour QCI field test.

Residential Energy Dynamics

http://www.residentialenergydynamics.com/

RED is best known for its tools for calculating the ASHRAE ventilation requirements. But the site goes way beyond that. There are some wonderful free tools here for zonal pressure diagnostics, calculating the R value of a wall using an infrared camera, design infiltration, using a pitot tube to measure airflow through a duct, etc. This is a gem!

Energy OutWest Weatherization Field Guide

http://www.energyoutwest.org/eow_library/__presentations_docs/_policy_specifications/EOWFieldGuide.pdf

A great field guide resource that was presented at the Energy OutWest Conference. (Great conferences to go to if you want to learn a lot about a lot of stuff.)

Weatherization Program Notices for 2015

http://bit.ly/1ekpiEp

These notices provide specific guidance on the DOE weatherization program. They are updated and added to on a continual basis, and they should be reviewed as part of preparing to take the QCI exam.

Index

Paul H. Raymer

Paul H. Raymer has been working with building science for more than thirty-eight years. He has developed and brought to market more than twenty products. He has taught building science courses to diverse audiences from home inspectors and weatherization specialists to architects and engineers.

Special thanks to Rana Belshe, Bob Davis, Tim Lenahan, Jeffrey Granger, and all the others who helped me to edit this print version of the book.

www.ingramcontent.com/pod-product-compliance
Lightning Source LLC
LaVergne TN
LVHW082352200926
848481LV00012B/3545